A GUIDE TO UNDERSTANDING LAND SURVEYS

A GUIDE TO UNDERSTANDING LAND SURVEYS

Second Edition

STEPHEN V. ESTOPINAL, P.E., P.L.S.

JOHN WILEY & SONS, INC.

New York / Chichester / Brisbane / Toronto / Singapore

In recognition of the importance of preserving what has been
written, it is a policy of John Wiley & Sons, Inc., to have books
of enduring value published in the United States printed on
acid-free paper, and we exert our best efforts to that end.

This publication is designed to provide accurate and
authoritative information in regard to the subject
matter covered. It is sold with the understanding that
the publisher is not engaged in rendering legal, accounting,
or other professional services. If legal advice or other expert
assistance is required, the services of a competent
professional person should be sought. *From a Declaration
of Principles jointly adopted by a Committee of the
American Bar Association and a Committee of Publishers.*

Library of Congress Cataloging-in-Publication Data:
Estopinal, Stephen V. (Stephen Vincent)
    A guide to understanding land surveys / Stephen V. Estopinal.—
2nd ed.
        p.   cm.
    Includes bibliographical references and index.
    ISBN 0-471-57717-0 (cloth)
    1. Surveying—Handbooks, manuals, etc.   I. Title.
TA551.E88   1992
526.9—dc20   92-10963

Printed in the United States of America

10 9 8 7 6 5 4 3

For my father, Eugene

FOREWORD

Like that of other highly skilled professions, the complexities of surveying are not well understood by the average person. Although common knowledge includes measuring and drawing maps, there is considerably more responsibility attached to a routine survey. In addition, the land surveyor is involved with many other things, such as the application of legal rules in deciding boundary questions, writing and interpreting land descriptions, researching records, resolving conflicts and encroachments, using aerial photography and satellite data, as well as testifying in court. Also important is the ability to work and communicate with others to address their needs or to satisfy their requirements. To have an appreciation for, and an understanding of, what a surveyor is concerned with requires some study.

Stephen Estopinal has ably explained in a readable manner many of the functions and procedures required in every survey—routine or otherwise. He has gone beyond that, however, and has delved into the reasons that make these functions and procedures necessary. With a basic understanding of what is done and how it is accomplished, the reader not only will have an appreciation for a survey, a plat, or a land description but also will be able to evaluate it in its proper perspective and realize any inherent inadequacies or discrepancies that may exist. Additionally, and perhaps even more important, the reader will have a much better idea of when a survey is needed to solve a problem or to obtain an approval, and what data will need to be collected or evaluated.

As Stephen Estopinal has outlined in Chapter 1, one of the objectives of this book is to meet the needs of others dealing with rights and interests in land. A highly positive aspect of this book is the author's continual relating of the survey and title aspects to one another, demonstrating that, although they are different, title and location are two separate entities, which are companions resulting in a single concept of land ownership and possession.

DONALD A. WILSON

Land Boundary Consultant
March 1992

PREFACE

The user of land information has been severely handicapped, because the survey profession had assumed that the information produced in maps and plats was as easily understood by the lay person as by the surveyor. Misconceptions and incorrect assumptions on the part of attorneys, landowners, realtors, and others who use the information produced by land surveyors have led to expensive and totally unnecessary paperwork, delays, and even litigation.

After being hired as a consultant on a case that proved to be one of unnecessary turmoil created by the misinterpretation of old survey information, I began to search for a good reference book that I could recommend to nonsurveyors, particularly attorneys, who frequently reviewed survey information. I discovered that libraries are filled with books about surveying or the science of measurement written for surveyors, as well as books about real property law written for lawyers and even books about real property law written for surveyors. Nowhere could I find a book about surveying that was written for the nonsurveyor. The first edition of *A Guide to Understanding Land Surveys* was my response to that need.

Since it was first published in 1989, I have received many comments on the book. Surveyors were delighted that they finally had a reference book that they could give to their clients that would help both to communicate. Attorneys, realtors, and other users of survey information were delighted that they finally had a book that they could read without wading through a bunch of "how tos" on the techniques of surveying. Each group found elements that they thought could be added or expanded on and areas that they thought could be trimmed down. This second edition is a response to that input, yet retains the unique character of the original work.

Chalmette, Louisiana
June 1992

STEPHEN V. ESTOPINAL

CONTENTS

7.   LAND RECORD SYSTEMS   77

A GUIDE TO UNDERSTANDING LAND SURVEYS

CHAPTER 1

PURPOSE OF THE BOOK

This book was developed to meet the needs of attorneys, abstractors, realtors, land planners, entrepreneurs, and others who must use and evaluate the work of professional surveyors. Boundary survey plats or maps, property maps, topographical maps, survey reports, and property descriptions, among other things, are all products of the professional surveyor. Rapid increases in technology, land values, and community planning, and the modern propensity for litigation have meant that more and more nontechnical individuals have begun to use and rely on increasingly complex and technical land information.

Persons who find themselves in need of the services of a professional surveyor may have difficulty in communicating to the surveyor just what it is they need. From attorneys to private citizens, the word "survey" has many different meanings. For example, a determination of boundary location for the purpose of erecting a fence is a "survey." The determination of the elevation of a building for the purposes of obtaining flood insurance is a "survey." The recovery or remonumentation of the boundaries of a parcel of land for the purpose of an exchange of title is a "survey." Each of these examples is called a "survey," yet the scope of work, the responsibilities of the surveyor, and the *expectations* of the person ordering the survey are all very different.

A better basic understanding of just what a surveyor does and does not do can make the difference between complete service and unsatisfactory results. Much litigation, confusion, and aggravation can be traced to a misunderstanding between the surveyor and the client. It is essential that the surveyor be informed of the exact purpose of the survey being requested. The completeness of the surveyor's report, the extent of his or her research, the areas examined, and other pertinent work vary greatly, depending upon the type of survey that is being conducted.

Compounding the problem is the fact that the person requesting the services of a professional land surveyor may not be the person using and interpreting the results of the work. This "third-party" use of a surveyor's maps or reports is rife with dangers, not the least of which happens when the user assumes that the surveyor did what the user wanted done when in reality the surveyor carried out the client's orders. For example, an owner of a vacant urban lot may request that a surveyor only recover the boundaries of that lot. Easements, servitudes, building restrictions, and other important title restrictions that impact on the enjoyment of that lot do not affect the boundaries of that lot. The surveyor would then recover or mark the boundaries and issue a report in the form of a drawing to the owner, showing the dimensions of the lot and the locations of the boundary markers. If, at a later date, the owner were to sell the lot, he or she could present this drawing to a buyer. The buyer, having no knowledge of the limited request of the previous owner, might then rely on the drawing produced as if it reflected all of the information about easements, servitudes, buildings, and other facts important to the buyer. The resulting lawsuit would charge that the surveyor neglected to perform the extensive work that the buyer needed when he or she complied with the limited request of the owner.

The grief, aggravation, and expense resulting from the previous scenario could easily be avoided if the buyer or the buyer's attorney confirmed that the drawing presented to them contained all of the information that they wanted to know. Third-party situations like this one have resulted in so many lawsuits that many surveyors now place explanations of the extent of the work done by them in producing the survey plat directly on the plat. Some state registration boards have even adopted a regulation or "standards of practice" that require such statements.

The land surveyor is an investigator, a detective more than anything else. The subject of the investigation is the location of the boundaries of real property. In the pursuit of the evidence necessary to determine, with relative certainty, the original location of a particular boundary line, the same rules of evidence apply as in any civil court. Surveyors not only measure angles and distances but also perform extensive records research in private and public files in an effort to reveal as much information as possible about the *location of boundaries*.

This extensive research does not normally include many other factors involved in property rights. Title insurance is a different service from a boundary survey, although most title insurance policies include the requirement that a boundary survey be performed. Easements, servitudes, building restrictions, setback, or side clearance for new construction, flood zones, and regulatory zoning, among other things that limit the use and enjoyment of real property, are not essential to the recovery of boundaries. If you wish to have all, or any, of these items shown on a survey plat, you must tell your surveyor.

The surveyor will report evidence of these factors limiting the free enjoyment of a real property parcel whenever it is discovered in the normal course of boundary recovery. If an original subdivision plat includes an easement, the surveyor will normally report it. If an easement is granted separately from the subdivision recordation, the surveyor may not have the occasion to discover it. The records search by a land surveyor takes a much different route than that of a title examiner. Provide the surveyor with the complete title record if you wish it to be shown on the survey plat. Otherwise, a boundary survey will result in the recovery and a report (plat) on the real property limits of a particular parcel, *and nothing else.*

This book is intended to bridge the ever-widening gap between the users of land boundary information and the producers of that information—professional land surveyors. To that end, the information in this book will be only as technical as necessary and will be presented in such a way that every reader, from one experienced in the use of land boundary information to the novice, will gain a better understanding of the profession of land surveying and the products of that profession. The early chapters of this book provide a very important background to understanding real property surveys, because they deal with the root causes of confusion and misunderstandings concerning boundary surveys and the survey plats or maps that are produced as a result of surveys.

Real property laws, traditions, and practices vary throughout the United States. This book will not attempt to address the specific details of every state but will describe in a general setting the common facts found throughout the United States. The reader is advised to determine how the particulars of real property laws and the principles and practices of boundary surveying in his or her area vary from the general concepts addressed in this book. Most local surveyors or professional surveying organizations are very happy to assist anyone in familiarizing themselves with local practices.

A large portion of this book is devoted to the writing, reading, and interpretation of legal descriptions of real property parcels. That portion will be devoted to the mechanics of legal descriptions and will not presume to advise the user on the appropriate form of legal description to use for a particular situation but, rather, will tutor the reader on the geometric and semantic aspects of describing real property parcels. The great differences between those regions of the United States known as "metes and bounds" areas and United States Public Land Survey (USPLS) areas will be explored in depth.

This book will serve both as a handy, daily reference guide and as a tutorial text. The reader may want to refer to portions of this book when ordering survey work or when evaluating survey plats, legal descriptions, or reports already received. Although this book will improve the reader's understanding of land surveys, it will not make the reader a land surveyor. Many highly technical aspects of measurements, computations, details, and equipment are only lightly referred to, if at all. Sophisticated research and recovery techniques, so vital to the recovery of boundaries, would make a book in itself. Your professional surveyor is the expert you must rely upon for these skills.

This book will not serve as a substitute for qualified legal assistance in the interpretation of real property rights or transfers. The survey plats and problems used as examples in this book are simplified versions of their more complicated actual counterparts. The laws concerning boundaries and real property rights differ from state to state and change over time. Some general rules are mentioned, but these are not to be interpreted as universal laws. The attorney is the professional who should be relied upon for the status of real property law in a particular area. The reader should make note of such regional variations as may exist when he or she is using this book.

CHAPTER 2

REAL PROPERTY ON A ROUND PLANET

2.1. OWNERSHIP

Land cannot be owned; at least, not in the same sense as physical objects are owned. People cannot carry their land with them to new locations. Land cannot be manufactured or destroyed. If one were to excavate a great hole in a property parcel, the land might be rendered useless but the real property parcel would still exist. "Reclamation" of submerged land is not manufacturing land but simply altering the surface condition of an existing area. The "sale" of a real property parcel does not result in a giant cookie-cutter stamping out of a hunk of land to be turned over to the purchaser. One cannot tour the country collecting scraps of land and assemble a ranch in California.

Real property ownership is not the ownership of land; it is the ownership of rights. The right to habitat, to cultivate, to deny access to others, and countless other rights are what is owned. All of these rights are intangibles and flow from the "sovereign" or governing authority. It is the armed forces of a national government that establishes sovereignty and defends the laws that dictate which rights may be acquired by individuals in a certain area. Not all national governments allow personal real property rights.

2.2. TITLE

The Constitution, laws, and statutes of the United States of America and the individual states allow the assignment of "real property" rights to private persons. Those persons entitled, under the law, to exercise the rights of real property ownership are said to "hold title" to a land parcel. It is these rights that are transferred in the sale of "land." Persons may gain "title" to real property through written and unwritten transfers. However, only recorded, written title transfers are merchantable.

2.3. BOUNDARIES

Real property boundaries are the invisible limits of the rights of real property. Tangible possessions have clear limits. Unless there has been a terrible accident, a person has no difficulty discerning where her or his car ends and another's begins. A person can exit the car without entering another's. This is not the case with real property. The limits of the rights of ownership cannot be seen. Physical evidence as to the location of that limit may be obvious, but the limit itself is an invisible and dimensionless concept. One must be on one side of a property boundary or the other. As soon as one crosses the limit of one set of rights, another set of rights is entered.

2.4. DEED

A deed is the written instrument that describes the assignment of the rights that are "owned" and the geographical limits of that assignment. A deed is not the same as a title, although the words are often used interchangeably. Although the limits of the rights of real property ownership are invisible, they must be identifiable. Every deed must communicate what rights are owned and where the boundaries or limits of those rights are. It is folly to attempt to transfer rights in real property without adequately identifying and communicating the location of the limits of the rights conveyed. The charge that our society places upon the professional land surveyor is to *recover, monument, quantify, and document the imperceptible limits of the intangible rights that constitute real property ownership.*

2.5. IDENTIFYING BOUNDARIES

There has always existed the need to identify and to communicate locations on the surface of the earth. Stone Age men and women could only relate to localities by naming mountains, rivers, or other features of the terrain and then specifying a locality as being near a certain feature. This first use of landmarks required that the receiver of the information be as knowledgeable about the terrain as the giver. Although this was adequate for identifying general areas to a person who was familiar with the territory, the advent of long-range travel, especially sea travel, required a method of identifying locations that did not rely exclusively upon terrain features.

As the earth spins in orbit around the sun, its motion can easily be detected by any observer using no other instrument than the human eye. Unlike the moon or the sun, the stars clearly reveal the constant, circular motion of the planet. It was this motion that led to the development of today's worldwide system of identifying points on the surface of the earth. In spite of the fact that some of the ancient observers may have even believed that the world was flat, the direction to the center of the apparent circular celestial motion became the standard direction of reference that we call "North." This standard remains to this day, simply because it can be easily and precisely determined anywhere on earth.

2.5.1. Latitude

Ancient seafaring societies noticed that the motion of the stars was not only circular but also regular and predictable. One of the earliest observations was that particular stars were known to pass directly over the same place every night. Many societies came to identify stars by the earth locations or "latitudes" that the stars would seem to track. The Polynesians, to this day, can identify a certain star by which island or island group that star passes directly over. The reverse is also true: A certain location can be identified by the star that passes overhead.

In reality, the apparent motion of the stars is the result of the spinning of the earth. The apparent center of rotation that the ancients observed is an indication of this axis of rotation. "North" is so universally detectable that it was inevitable that the system of identifying locations that developed used this axis as the reference base. Sketch 1 shows how the concept of "latitude" first developed. Note that the angle measured to determine the latitude of the observer's location is actually the angle between a horizontal line tangent to the level surface of the earth and the axis of rotation. Points on the surface at which the same angle or latitude is observed can be connected to form a curved line circling the earth.

This connection of points on the surface of the earth of equal latitude are called "latitude lines." By definition, latitude lines are east–west lines, and, as shall be shown later, of all the latitude lines, only the equator (zero degrees latitude) is a "straight" line. The latitude of a particular point can be used as a measure of the distance north or south of the equator and is so easily measured that early explorers often sailed north or south to the latitude of their destination and then east or west to reach shore. This conve-

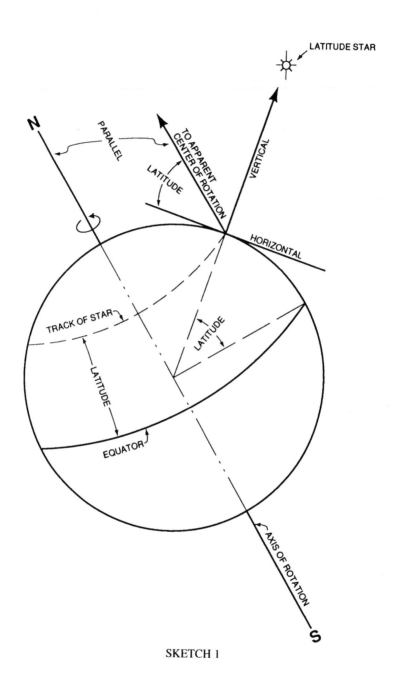

SKETCH 1

nient measure of north–south distances often caused early maps to be quite accurate in reporting latitudes but woefully inaccurate in reporting east–west dimensions.

2.5.2. Longitude

The axis of the earth's rotation is, therefore, a naturally occurring reference that can define directions on the surface of the earth, as well as distances in a north–south direction. In order to identify a particular point along a line of latitude, the concept of "longitude" (shown in Sketch 2) was developed.

Just as latitudes are a contiguous series of points with a common factor (the measure of the angle between the earth's axis and the horizon), longitudes are a series of contiguous points with the common factor of being aligned along the apparent axis of rotation. These north–south lines are sometimes referred to as "meridians" when they are used to define other lines or directions. Unlike latitudes, the reference for longitudes had to be arbitrarily chosen. Zero degrees longitude is defined as passing through a particular point in Greenwich, England, and is called the "prime meridian." Longitudes are used to define or to measure distances in an east–west direction. The combination of a north–south distance and an east–west distance can be used to define a particular point on the surface of the earth. Theoretically, only one pair of latitude and longitude values can be associated with a specific surface point.

Although longitudes are reported in degrees, just as are latitudes, time is the factor actually measured by observers in determining the longitude of a particular location. In order to accomplish this measurement, an observer might note, for instance, that a particular star is directly over the prime meridian at a certain time and that that same star is directly over the longitude of the observer some time after that. The elapsed time is directly related to the distance from the prime meridian to the longitude of the observer. The earth completes approximately one rotation (360 degrees) in 24 hours. Therefore, the ratio of elapsed time to 24 hours is the same as the ratio of degrees longitude to 360 degrees. In actual practice, the stars observed, angles measured, and elapsed times noted are never as direct as this simple example.

Latitudes and longitudes are reported as if they were angles measured at a point at the center of the earth, when, in fact, the measurements are of time and of angles made at the surface of the earth. This would not cause a major distortion if the earth were a sphere with no irregularities of surface or gravity. But precisely because the earth is not a sphere and because the observations are made on the surface of the earth, the shape of the earth and

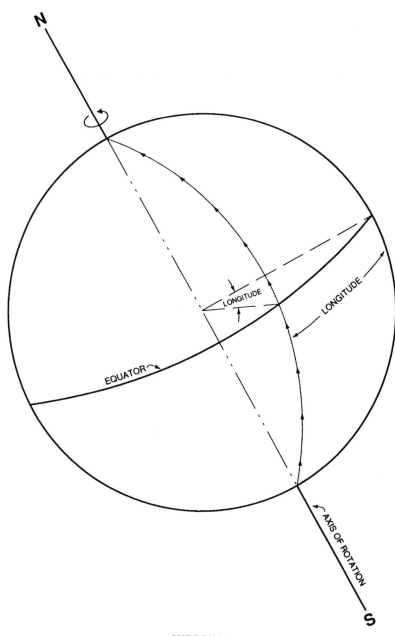

SKETCH 2

the distortions caused by the irregularities of the surface must be understood in detail.

2.5.3. Astronomic Position

With the first attempt to account for the irregularities of the earth's surface, a method was developed by which all measurements or observations were adjusted to account for elevation above sea level. This method developed a theoretical surface that covered the earth and that was everywhere perpendicular to the pull of gravity. The theory behind this assumed that, if the entire planet were covered with water, the planet would be a perfect sphere. The surface that resulted from this "reduction to sea level" is called a "geoid." The latitude and longitude based upon celestial observations used to identify a specific location are called that location's "astronomic position." Astronomic positions are computed based upon the geoid surface. Unfortunately, a geoid is not a sphere, nor is it a regular surface.

Gradually, the realization developed that the anomalies of the gravitational forces around the world produced a geoid that was "dimpled" and undulating. Astronomic positions are, necessarily, based upon the pull of gravity. Astronomic latitudes and longitudes are not, therefore, perfectly circular curves or perfectly straight lines on the geoid surface. The positions reported by astronomic observations are distorted by the irregularities of the geoid.

Astronomic positions were quite adequate until the methods and precision of measurement became so refined that the irregularities of the earth's shape became detectable. Astronomic positions lack the precise predictability necessary for accurate navigation and mapping. The science of geodesy, the study of the size and shape of the planet earth, and the science of cartography, the study of mapmaking, have developed methods to reduce the uncertainty inherent in astronomic positioning.

2.5.4. Geodetic Position

The need to compute precise distances between locations of known positions or to compute the position of a location based upon distance and direction from a known position requires a regular, mathematically identifiable surface on which computations can be made. The geoid can be approximated by a theoretical shape or model based upon regular geometrical formulae. The mathematical shape or model chosen for this approximation is called a "spheroid" or "ellipsoid" and forms a theoretical surface on which the values of latitude and longitude are computed for any earth location.

The latitude and longitude of a location based upon an ellipsoid is called the "geodetic position." Imagine that the surface of the earth is transparent and that the surfaces known as the ellipsoid and the geoid are glass bulbs around the center of the earth. A man standing on a high mountain could shine a light straight down. Where that light hit the geoid would be the astronomic position of the man. Where that light hit the ellipsoid would be his geodetic position.

The geodetic position is what is reported for navigational and mapping purposes. For most localities and for many purposes, the differences between astronomic and geodetic positions are not significant. The ellipsoid used in a particular map or in a particular country is usually defined by national or regional statutes. The standards for the United States of America and each of the member states will be discussed at length later in this book.[1]

Locations are reported by a longitude west or east of the prime meridian and a latitude north or south of the equator. Longitudes are always less than or equal to 180 degrees. Latitudes are always less than or equal to 90 degrees. The values reported are the apparent distances along the surface of the geoid, in the case of astronomic positions, or the ellipsoid, in the case of geodetic positions.

2.5.5. Flat-Earthers

The spherical reference systems of astronomic and geodetic latitude and longitude are perfectly applicable to defining locations on a planet or, more precisely, on the surface of a spheroid, as shown in Sketch 3. On the other hand, the basic instincts of the human mind and the limited input of our senses tell us that the world is flat. Because the earth is so large compared to the range of our senses, we conceive of the surface of the earth as being two-dimensional. This concept of a two-dimensional surface is reinforced by our use of maps and our insistence on using two-dimensional geometric terms that describe features on the map when we are referring to features that are on the surface of the earth.

Consider the following trip. A woman on a Montana ranch begins a trip by walking due north for exactly 1 mile, then turns right and walks exactly 1 mile due east, turns right and walks exactly 1 mile due south, and then turns right and walks exactly 1 mile due west and stops. The question is, "Where is the woman in relation to her starting point?" If you say that she has returned to her starting point, then welcome to the flat-earth society!

Even after sketching such a trip out on paper, our instincts blind us to the fact that the trip took place on the curved surface of the earth and not a flat surface. The sketch on a sheet of paper is a two-dimensional representation

---

[1]See section 5.6: Advantages of State Plane Projections.

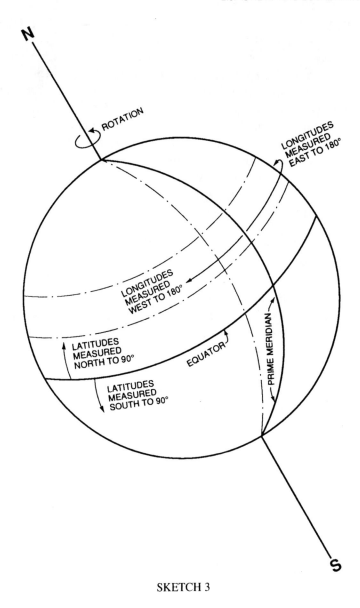

SKETCH 3

of a three-dimensional event. In spite of discoveries, observations, knowledge, and measurements to the contrary, each of us persists in thinking of the surface of the earth as flat. This "flat-earth" mentality must be overcome in order to begin to understand land surveys and the maps or plats that represent these surveys.

All of this would be academic to most of us if it were not for the concept of ownership. The law recognizes the fact that persons can claim, as personal property, certain rights to portions of the planet. Real property law is that body of law devoted to defining the limits of these "rights of ownership" over portions of the earth. Real property boundaries are imaginary limits intended to define the extent and dimensions of such ownership. This ownership is limited neither to the surface of an imaginary spheroid nor to the surface of the earth. The boundaries of real property rights or ownership are imaginary limits created by law in an attempt to define the actual, corporeal use of the land. The professional land surveyor serves as the bridge between the physical limits of real property and the imaginary concepts of ownership boundaries. The great value often attached to continued enjoyment of real property rights demands *permanence* and *recoverability* of the physical limits of real property boundaries.

2.6. CORNERS

It is possible to mark a parcel that has straight land line segments on all sides. It is also possible for each of the four sides to be 1 mile in length. Instead of defining the boundaries of the parcel by marking each step of a walking man, the end points of each side could be marked and the boundary could be defined as a straight land line *between the end land points or "corners."*

The desire for regular shapes, parallel lines, and constant directions for straight lines is resolved when the boundaries are so defined. For this reason, as well as others to be discussed later, the vast majority of boundaries in the United States are defined by the location of corners.

CHAPTER 3

GEOMETRY

In order to better understand exactly what it is that professional land survey-ors are presenting in their plats, maps, and reports, a closer look at the somewhat esoteric concepts of geometry and geodesy is necessary. Some of the words used in describing boundaries are common geometric terms and have generally accepted definitions that are *very different* from what is meant when these terms are used in real property situations. Geometry is consistent in that the terms used can describe two- or three-dimensional concepts. Maps are two-dimensional drawings, and the geometric terms describing map fea-tures are consistently confused with the three-dimensional surface features that the maps represent.

3.1. PLANE GEOMETRY

Consider these basic definitions of geometric terms:

Space—An extent or continuum in which objects can exist and have rel-ative positions and directions.

Point—A specific, dimensionless location in space having a unique po-sition.

Line—A series of contiguous points such that any point in the line is directly adjacent to two and only two other points in the line.

Straight Line—Any line in which the route of measure of the shortest distance between any two points on the line lies entirely within the line.

Ray—A portion of a straight line originating or emanating from a point.

Line Segment—A portion of a line beginning at one point and ending at another.

Distance—A measure of the separation of two points.

Length—The sum of all of the distances between consecutive points of a line segment.

Surface—Any array of contiguous points, having no thickness, that completely separates two distinct spaces.

Plane—A surface such that the shortest route between any two points on the surface lies entirely within the surface.

Area—The measure of a bounded surface.

Volume—The measure of a bounded space.

Angle—A measure of the relationship of two lines, rays, or line segments.

If one were to restrict the concepts of geometry to the two dimensions of a flat plane, then the three-dimensional terms of space and volume would be eliminated. The concept of lines also changes subtly. Straight lines in a two-dimensional system either intersect or are parallel. The concept of angles is rendered much simpler, and the entire system can be modeled quite easily by sketching on paper. There are also one-dimensional systems that are frequently used but rarely recognized as such. One particular one-dimensional system, the deed description, will be discussed at length in Chapter 11.

It is important to keep in mind that, in spite of the fact that one- and two-dimensional systems exist, our world is three-dimensional, so, therefore, whenever real property is defined or exchanged, a three-dimensional portion of our planet is involved. This fact requires that the geometric definitions already given be redefined to reflect what is really meant when these terms are used in describing real property. In order to assist the reader, when terms are used that may confuse the real property definitions and the geometric definitions, the use of the adjective ''land'' will emphasize the real property definition—for example, ''land line,'' ''land boundary line,'' ''land point,'' and so on when the real property definition is intended. In everyday practice, the adjective ''land'' is not used.

Anyone who has dealt with real property boundaries, title transfers, or surveys may have noticed that some of the words defined earlier in this chapter concerning dimensional geometry are often used in describing real property. These geometric terms, when used in real property descriptions, are misused in a curious mixing of two-dimensional and three-dimensional concepts. It is very important to understand how these words are misused, or, more accurately, redefined in real property descriptions before an appreciation of just what a boundary survey really is can be reached. The distinctions of what is conveyed, what is physically marked, what is drawn on a survey plat, and what is described in legal instruments are vitally important.

3.2. LAND POINT

An easy definition of a "land point" as used in real property descriptions would be something similar to "a specific location on the surface of the earth." However, it is much more than that. For instance, if a property corner were located on the side of a hill and the hill were removed by earthwork, the property corner or land point would remain. Its location would not be thought of as having "moved," although the surface of the earth would now be several feet lower than the original surface. If, on the other hand, a property corner were located in a swamp or a low spot and the land were later filled or leveled, the location of the property corner would not be altered as far as the real property definition of that parcel was concerned. Changes in elevation do not alter the locations of land points.

Sketch 4 illustrates a real property land point that was marked (monumented) on the earth's surface in 1802. Years of erosion and earthwork have destroyed the monument and have lowered the surface several feet. In 1980, the land point was again monumented at a different location in space, several feet nearer the center of the earth, yet the location of the land point did not change when it was defined in land surveying terms.

What is called a "point" in real property terms is, in geometric terms, a ray emanating from the center of the earth and extending through the surface of the earth, where it is usually marked or defined, and beyond the earth as far as the legal concepts of "air space" and "reasonable control" extend. Real property descriptions call this ray a "point" because humankind's flat-earth instincts and, more important, the surveyor's plat depict the real property parcel as being a plane.

3.3. LAND LINE

In real property descriptions, a "land line" is a series of land points. There are many different kinds of land lines, but the land line most commonly

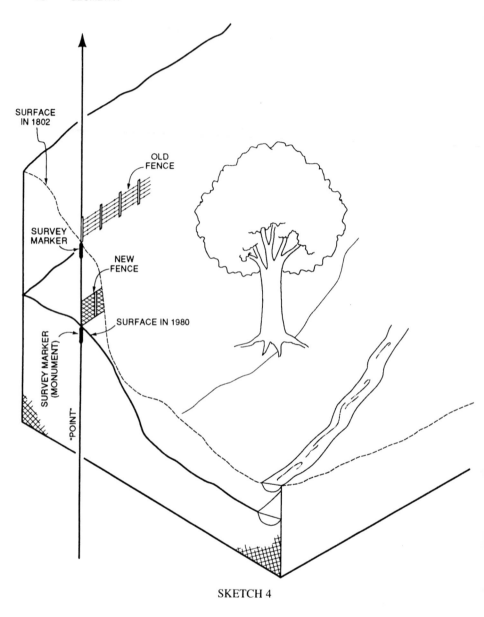

SKETCH 4

encountered by the lay person is the boundary or property line. Boundary or property land lines define the limits of rights or possession of land areas. Like real property land points, property land lines are not thought of as "moved" or "shifted" if changes of elevation take place. Land lines are imaginary or theoretical limits that have no thickness but do have direction and location. Because land lines are a series of land points, the geometric description of a land line is a vertical surface containing the center of the earth.

3.4. STRAIGHT LAND LINE

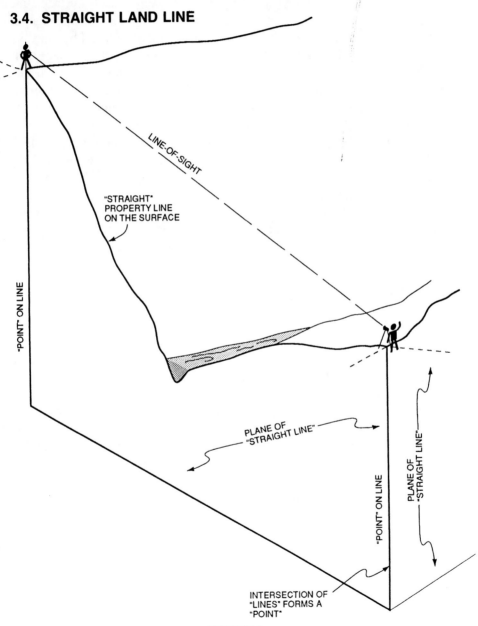

LINE-OF-SIGHT

"STRAIGHT"
PROPERTY LINE
ON THE SURFACE

"POINT" ON LINE

PLANE OF
"STRAIGHT LINE"

"POINT" ON LINE

PLANE OF
"STRAIGHT LINE"

INTERSECTION OF
"LINES" FORMS A
"POINT"

SKETCH 5

The direct, line-of-sight, route between two distinct locations (land points) on the surface of the earth has been traditionally defined as a straight land line. Historically, straight land lines are marked on the ground by aligning all interior points of the land line between the end land points. Because land points are geometrically rays and land lines are geometrically surfaces, the procedure for establishing a straight land line results in the establishment of a *flat* vertical plane passing through the center of the earth and bounded by the rays that make up the end land points. Therefore, a straight land line is not altered by shifting terrain. A land line may travel up hills, down into valleys, and across rifts in the surface and still be straight. Any ray from the center of the earth that is contained in the plane of a straight line or in the surface of any land line is said to be a "point on the line."

Note that, in Sketch 5, the surveyor's instrumentman has set up her theodolite such that the land point passes through, and is coincidental with, the axis of rotation of the instrument (in surveying jargon, this is "occupying a point") and is directing the rodman, who is holding a vertical pole (called a range pole), to a "point on line." The vertical pole will be "on line" when the ray that is the land point passes through, and is coincidental with, the vertical range pole.

3.5. PLUMB LINE

In the real property definitions of both a point and a line, the concept of "up" and "down" or "above" and "below" were used. Many children's programs deal with these concepts by simple demonstrations. Closer examinations of these instinctive terms have revealed subtle complexities, some of which need to be addressed now.

"Up" is away from the pull of gravity. "Down" is toward the pull of gravity. The pull of gravity is *not* always directly toward the center of the earth but toward the apparent center of gravitational attraction. A line that is everywhere aligned with the pull of gravity is said to be a "plumb line." For most localities on the surface of the earth, the pull of gravity is very nearly toward the center of the earth. The effects of the difference is real, measurable, and, in some rare cases, affects the definitions of the location of real property lines.

3.6. LEVEL

A level surface is a surface such that each and every plumb line passing through that surface does so perpendicularly. The geoid discussed earlier is a level surface. "Flat as a mill pond" is a contradictory phrase. On a per-

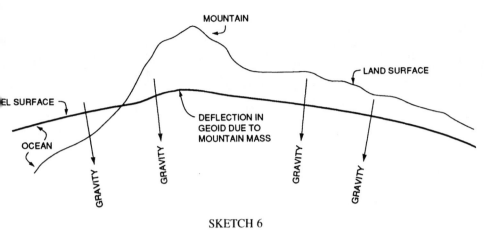

SKETCH 6

fectly calm day, a mill pond is level, not flat. Notice that, in Sketch 6, the level surface ripples or undulates as deflections in the pull of gravity are influenced by irregularities in the earth's land masses.

3.7. LAND DISTANCES

Now that we see that real property points are geometrically rays and real property lines are geometrically surfaces (a "straight line" being a flat plane), let us examine how "distances"—that is, the separation of two land points—are defined. In geometry, a "distance" is the measure of the separation of two points in space. In Sketch 7, it would be geometrically correct, for example, to define the distance between geometric point "A" on the surface of a mountain and geometric point "B" at the foot of that mountain as the measure of geometric line segment having "A" and "B" as end points. This is not the real property definition of the distance between these land points, because real property land points are actually rays.

Although the separation of two rays can be measured by the angle between them, and some property and many national boundaries are measured in this way, in real property terms, the distance between "A" and "B" is defined as being measured along a level surface from the intersection of one ray to the intersection of the other. Land distances are often referred to as the "horizontal" distance.

Notice that, under the real property definition of "distance," changes in the topography do not alter the land distances. The mountain could be scraped away, but the land distance between point "A" and point "B" would remain the same. It is because of the need to reduce field measurements,

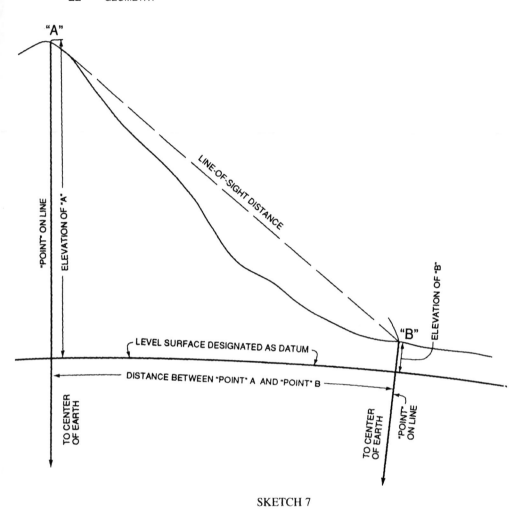

SKETCH 7

which are geometric distances, to land distances, as well as the need to apply a multitude of other correction factors, that, except for very short distances, it is rare for the values shown on any survey plat to be the result of a single, direct measurement.

Distances are dependent upon a standardized length. There is no naturally occurring object that is consistently and universally the same length always and everywhere, so a standard had to be created. The meter is the international standard unit of distance measurement. Although it is being slowly replaced by the meter, the U.S. survey foot is the most common unit of measure for real property boundaries in the United States. The U.S. survey foot is slightly longer than the U.S. standard foot, the difference being only about three-thousandths of a foot in 1 mile. This difference is only significant in long-range, very high precision work. Other units of measure and the conversion factors for each are shown in Table 1 in the Appendix of Tables.

3.8. ELEVATION

Distances can also be used to describe or measure the separation of two level surfaces. This sort of measure is usually identified as "elevation." The most common use of elevation is in describing all surfaces from a common base or vertical datum. In Sketch 7, the level surface shown might represent such a datum. The distance from the datum to the level surface containing point "B," measured along a plumb line, would be the elevation of "B" as referenced to that datum.

Many different areas have established, through convention or municipal and state government, local vertical datums. These different datums are being rapidly replaced by the federal standard datum established by the National Geodetic Survey (N.G.S.). This federal datum is known as the national geodetic vertical datum (NGVD). NGVD is frequently and incorrectly called mean sea level (MSL). NGVD is not a sea level datum. Local mean sea level will vary greatly from NGVD, and NGS has consistently emphasized that NGVD is not a sea level datum. In spite of this, even federal agencies will occasionally refer to NGVD as "mean sea level" or "MSL"

Elevations are every bit as important in real property measurements as are horizontal distances and angles. More personal disasters have occurred because an individual's real property was below a certain elevation (a high-water mark, for example) than because of shortages in horizontal measure or improper horizontal location. In many regions of the country, there are significant changes in elevations taking place because of subsidence or heaving of the surface. Some changes in reported elevations are superficial—the result of improved measurement procedure. Many more of the changes are the result of actual surface movement. It is important, in regions of rapid changes, that elevation information be current.

3.9. LAND AREA

In the same way that real property distances are horizontal measurements usually defined at some datum, land areas are also defined as "horizontal" or "level" and are usually referenced to the same datum as the horizontal distances. The actual surface of a tract of ground—that is, the air–land interface, can be altered by earthwork. Leveling hills or plowing furrows alters the exposed surface of dirt, but land areas, as defined in real property terms, cannot be altered by earthwork.

3.10. HORIZONTAL ANGLES

Because real property lines are actually vertical planes, the intersection of these planes will form a vertical ray or a land point. A plane perpendicular to this vertical ray will also intersect with these vertical planes, forming two "horizontal" geometric lines. It is the angle between these two "horizontal" lines that is reported as the angle of a property boundary intersection.

Many have found it convenient to refer to the angle between two such lines as if the measurement were an action that took place by turning from one line to another. When this sort of "reckoning" is used, angles are said to be positive, or to the right, if the direction of the imaginary "turn" is

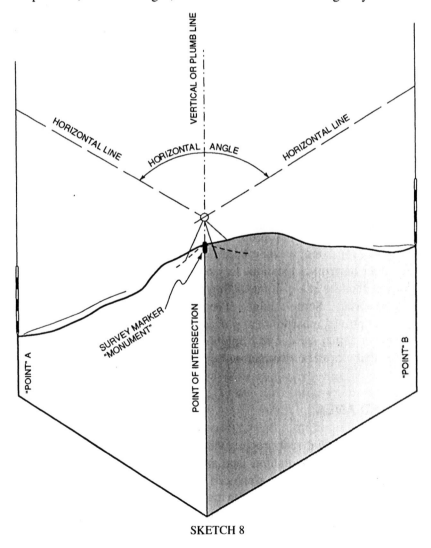

SKETCH 8

clockwise when observed from above, and negative, or to the left, if counterclockwise.[1] Consequently, the angle in Sketch 8 could be said to be 90 degrees, or to the right, from ''A'' to ''B,'' but −90 degrees, or to the left, from ''B'' to ''A.''

3.11. VERTICAL ANGLES

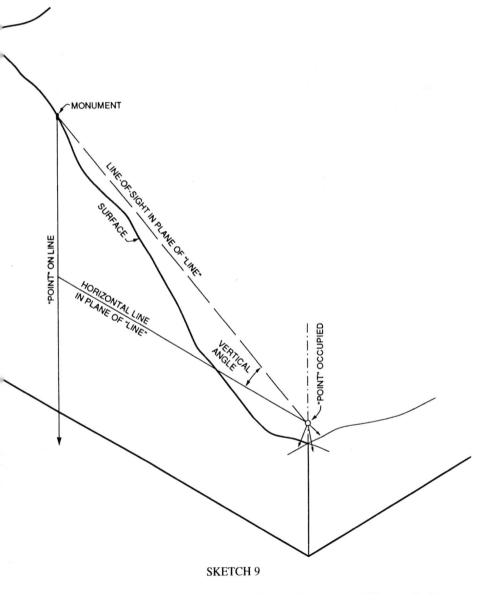

SKETCH 9

[1]The actual process of measuring angles is not at all dependent upon the ''direction'' of the reckoning.

Vertical angles are measurements between geometric lines in a vertical plane. Because real property boundaries are vertical planes, vertical angles rarely find their way into property descriptions. Vertical angles are commonly measured in order to compute horizontal and vertical distances. Sketch 9 illustrates a vertical angle.

3.12. DEGREES, MINUTES, AND SECONDS

All angles are based upon a definition of a unit, not a standard. A point traversing a circle will, when observed from the center, appear to travel the same distance to complete one circuit, no matter where that circle or observer is in the universe. One could contact an alien on another planet and describe a ''degree'' as one-three hundred-sixtieth of a circle, and that alien could duplicate the standard, having never seen an example. On the other hand, no amount of verbal description would allow the alien to duplicate the U.S. survey foot; the alien would have to be given a physical example in order to duplicate it. The relationship of the circumference with the radius of a circle is a natural constant. All angular measurement is defined in terms of a full circle. Table 2, in the Appendix of Tables, lists the most common definitions for angular measurements. In the United States, the most common definition of angular measurement is 360 degrees to a circle. The degree (noted by the symbol °), is divided into minutes (noted by the symbol ′), which is one-sixtieth of a degree. The minute is divided into seconds (noted by the symbol ″), which is one-sixtieth of a minute.

3.13. MAPS OR PLATS

A map or plat is a report of a survey in the form of a drawing. The root cause of the confusion between the terms used in real property descriptions and between identical terms, with very different geometric meanings, is the widespread use of maps or plats. Since the invention of paper, and perhaps even before that ancient time, people have communicated to one another the location of places on the surface of the earth by drawing lines on a flat surface. First as lines scratched in the dirt and then as ink lines on paper, the map was developed as a communication aid. Even after the concept emerged of drawing line lengths and angles to scale, the map or plat was, is now, and always will be a two-dimensional representation of a three-dimensional reality. A map is not a model. The interaction of ink lines drawn on a map is not the same as the interaction of real property land lines on the earth. Yet we persist in using the words and descriptions of the two-dimensional map as if it, and not the earth's surface, were the thing being described.

CHAPTER 4

DEFINING NORTH

Descriptions of real property boundaries often use terms that report the orientation of each of the boundary lines with a common line or datum. This orientation is called the "direction" of a boundary. The most common datum for boundary directions is "north." Once again, the commingling of three-dimensional terms, two-dimensional instincts, and one-dimensional verbiage makes it necessary to separate and explain each concept in terms of what is really happening, what is shown, and what is described.

No concept in real property boundaries is so widely misunderstood as the term "north." The misuse of this term in real property boundary definitions is so pervasive and so perfectly demonstrates the distortion of three-dimensional reality that a complete understanding of this single term and its many variations is required.

4.1. TRUE NORTH

"True north" is a three-dimensional term. In real property terms, straight land lines are defined as "true north" if the axis of rotation of the earth is entirely contained in the plane of that straight line. There are an infinite number of "true north" lines. All "true north" lines are straight land lines, but *none are parallel.*

4.2. ASTRONOMIC NORTH

"Astronomic north" was long believed to be true north. The apparent rotation of the stars, first observed by the ancients, is still used today as an accurate and universal method of determining the rotational axis of the earth. Unfortunately, the measurements of the angles between the horizon and the axis of rotation of the earth depend upon an accurate determination of "down." We must usually make the observations of the apparent movement of the stars at a point on the surface of the earth. This close proximity with the surface of the earth means that the pull of gravity, the "plumb line," will not always point toward the center of the earth. Land mass anomalies frequently deflect the plumb line from the vertical. Because these deflections distort celestial observations, astronomic north lines are not quite straight land lines but are meandering land lines that closely approximate true north. Astronomic north lines can be adjusted to "true north" by the application of regionally varying correction factors. These corrections are typically, but not always, very small.

4.3. MAGNETIC NORTH

"Magnetic north" is not directly connected to the rotational axis of the earth. The earth has a naturally occurring magnetic field focused on points near the arctic and antarctic regions. These points of focus, or poles, are not stationary, and the magnetic lines associated with these poles are far from straight and are not true land lines, for their location and intensities vary with altitude. These "lines of magnetic force" are easily interfered with or deformed by surrounding conditions. Alignment with these magnetic lines defines "magnetic north." The corrections required to translate magnetic north, as observed at a point on the surface of the earth, to true north vary so rapidly that it is quite useless to attempt even a moderately accurate determination of true north based upon magnetic observations. In spite of this, many old survey procedures were regularly based upon magnetic lines and required the surveyor to adjust the directions observed on a compass by what was believed to be the variation of magnetic north from astronomic north, called "declination," or sometimes "deviation." The results of such practices will be discussed later.

4.4. ASSUMED NORTH

When the direction depicted as north is based upon something other than observations intended to detect the axis of rotation of the earth, that direction is referred to as "assumed north." Many boundary surveys, if not most, use an assumed north as the direction of orientation. Many survey plats fail to state the type of "north" used to define the directions on the plat. In such cases, it is almost certain that the north used was an assumed north. It is also quite common for a survey plat, especially an old one, to report the base direction as "true north" when, in fact, the direction of orientation is an assumed north. *Assumed north lines are unique to each particular survey and have no fixed relationship to any other assumed north.* Chapter 9, which deals with the evaluation of survey plats, will demonstrate how to identify the type of "north" used in a survey plat.

4.5. GRID NORTH

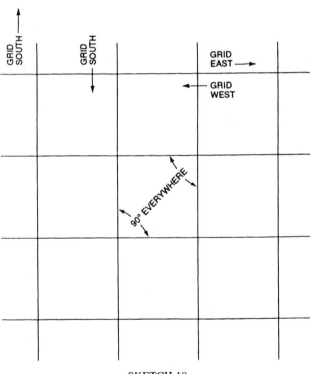

SKETCH 10

Of all the "north" terms discussed in this chapter, "grid north" is the only two-dimensional term. If a flat plane is divided into a set of evenly spaced parallel straight lines intersecting at right angles (90 degrees) with another set of evenly spaced parallel straight lines, the resulting pattern is a "grid."

"Grid north" could then be defined as a direction parallel with one of these sets.

This grid pattern (Sketch 10) could be defined as having a specific relationship to portions of the surface of the earth. When this occurs, such a grid pattern can become an effective tool in mapping the earth's surface. All maps, which are two-dimensional planes, depend in some way upon a grid pattern to control locations of the drawings and symbols that represent land features. The science of cartography is the highly complex and technical study of the methods and concepts necessary to present the surface of the earth as a two-dimensional grid (map). (Later in this book, we will examine briefly some of these methods.) Unlike true north, astronomic north, and magnetic north, grid north lines are parallel.

4.6. DIRECTIONS

The various concepts of "north" discussed so far have dealt with land lines or planes that are oriented along the earth's axis of rotation. There is an infinite group of land lines or planes that are not "north lines." These land lines can be, nevertheless, defined by their relationship to a particular "north." Azimuths and bearings are two universally accepted methods used to describe that relationship. Both of these methods are systematic reports of the angles formed between a land line and a meridian. The particular meridian used *must be defined.*

In the examples for azimuths as well as for bearings, the values report the relationship of land lines (flat planes) that intersect at a land point (vertical ray). The various definitions of direction are the same for each of the types of "north" involved. The methods of identifying directions and the verbiage employed are the same in two-dimensional systems.

Imagine a flat plane perpendicular to the line of intersection of a land line and a meridian. The vertical ray that is the line of intersection will appear as a geometric point on that imaginary plane. The vertical planes that are the land lines and the meridian will intersect the imaginary plane as geometric lines. The description of the relationship of the land line and the meridian has now been reduced to a two-dimensional image. This relationship is only valid *at a particular land point of intersection.* Reducing an intersection to a two-dimensional relationship for the purposes of describing the direction of a land line *does not eliminate the three-dimensional nature of the land lines involved.*

In Sketch 11, the line A–C intersects the north or meridian line at point B. Because the meridian is a straight line at point B and line A–C is a straight line at point B, then only one of the angles created needs to be

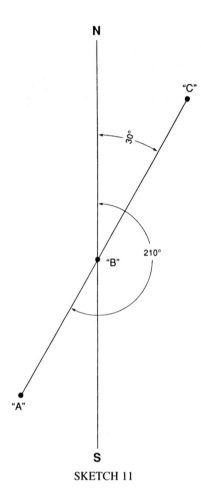

SKETCH 11

reported or measured to define the direction or orientation of line A–C. Because azimuths and bearings developed as an aid to navigation or travel, both have the ability to distinguish the direction "B" to "A" from "B" to "C." This one-dimensional aspect often causes confusion among users of land information.

4.7. AZIMUTHS

Azimuths are "measured" or "reckoned" from the north in most cases.[1] Some organizations "reckon" azimuths from the south. Azimuths are al-

[1]In actual practice, the measurement of angles involves a process that is independent of such "reckoning."

ways said to be clockwise, meaning that the observer would "turn right" from the meridian to the line being reported. If the observer is Sketch 11 were to face north, a right turn of 30 degrees would cause the observer to face "C." A right turn of 210 degrees (30 degrees + 180 degrees) from north would cause the observer to face "A." Therefore, line "A–C" could be reported as having an azimuth of 30 degrees or 210 degrees. *Both azimuths describe the same line.* Azimuths are always less than 360 degrees.

4.8. BEARINGS

Bearings are "reckoned" clockwise or counterclockwise[2] from the meridian and are always less than, or equal to, 90 degrees. In the case of bearings, the reporter defines lines as being east or west of north, or east or west of south. This method is easily comprehended and widely favored by most users of real property information.

In Sketch 12, land line "D–F" has been added to the previous sketch. If an observer at "B" were to face north, a right turn of 30 degrees would cause the observer to face "C." The bearing to "C" is north 30 degrees east. If that same observer at "B" were facing south, a right turn of 30 degrees would cause the observer to face "A." The bearing to "A," therefore, is south 30 degrees west. The bearing of land line "A–C" could be reported as being north 30 degrees east or as being south 30 degrees west. *Both bearings define the same land line.* The distinction of when to use one rather than the other will be discussed in Chapter 11.

If the observer were to face north and turn left (counterclockwise) 10 degrees 15 minutes 22 seconds, then the observer would be facing "D." The bearing to "D" is north 10 degrees 15 minutes 22 seconds west (written N 10° 15′ 22″ W). If the observer were to face south and turn left (counterclockwise) 10 degrees 15 minutes 22 seconds, then the observer would be facing "F." The bearing to point "F" is S 10° 15′ 22″ E. Both bearings describe land line "D–F."

If the values of the angles are the same, north–east bearings and south–west bearings describe the same line and are clockwise, or reckoned to the right. If the values of the angles are the same, north–west bearings and south–east bearings describe the same line and are counterclockwise, or reckoned to the left.

The concepts for bearings and azimuths discussed and sketched earlier are based upon a relationship of lines *at a particular point.* In two-

---

[2]"Clockwise" is defined as a turn to the right from the meridian, and "counterclockwise" is defined as a turn to the left.

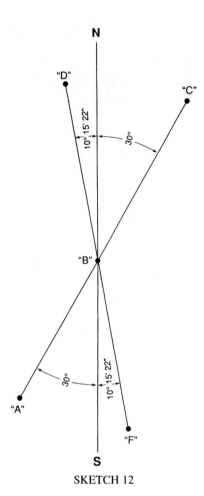

SKETCH 12

dimensional geometry, the meridian and all north–south grid lines are parallel, but, in the three-dimensional world of real property boundaries, meridian lines, such as longitudes, are not parallel with any other north–south line. Because straight real property lines are actually flat planes through the center of the earth, the following facts are true, although they may come as a shock to the "flat-earth" believers among us.

1. There cannot be two parallel straight lines. Because every straight land line is actually a plane containing the center of the earth, then every such plane must intersect. For any two straight lines, there are two places on the surface of the earth where these lines intersect; therefore, no two straight land lines can ever be parallel!

2. The bearing of straight lines (except for north–south lines and the

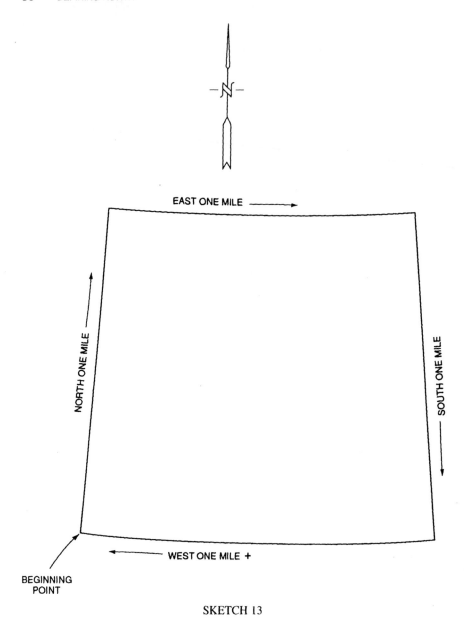

SKETCH 13

equator) vary throughout the line. Because none of the north–south lines are parallel, then a straight line that crosses several north–south lines must do so at a different angle for each north–south line.

A simple examination of a globe is all that is needed to confirm that north–south lines (lines of longitude) are not parallel (they meet at the poles).

East–west lines (lines of latitude), on the other hand, are parallel but, except for the equator, are not straight land lines.

The consequences of these facts in documenting real property transfers is best demonstrated by an example. Let us return to the ranch in Montana. We might attempt to mark a square mile of land in many ways. First, if we were to mark every step of the woman walking in the scenario, we would notice several things. The north leg of the trip would be a straight land line segment, 1 mile long. The east leg of the trip would be a land line segment curving to the north, 1 mile long. The south leg of the trip would be a straight land line segment, 1 mile long. The west leg of the trip would be a land line segment curving to the north, and, in order to return to the starting point, the walker would be required to travel slightly more than 1 mile.

In this example, such a closed circuit could represent a land parcel. This parcel could be described very accurately by reporting the directions and distances of the route. Sketch 13 is an exaggerated drawing of such a parcel in Montana.

Although the parcel shown in Sketch 13 is completely possible, the curved lines of the northern and southern boundaries are difficult to mark on the ground. By curving the lines on the sketch, the actual shape of the parcel can be approximated. This is not possible with larger areas of land. The convenience of straight boundary lines,[3] as well as the human desire for regular shapes, reduces the acceptance of such a strict geodetic definition of real property parcels.

---

[3]The greatest convenience of straight land lines is that only the ends (corners) need to be marked to define clearly the location of that line on the ground.

CHAPTER 5

PROJECTION SYSTEMS

5.1. PROJECTIONLESS MAPS

The first attempts at drawing maps in a scaled and controlled manner where the distances, locations, and directions, as observed on the ground, were reproduced in miniature on a sheet of paper failed miserably. Except for very small land parcels, these first attempts at mapping in this manner (now called "projectionless mapping") resulted in inconsistencies of ever-increasing magnitude as the area being mapped was enlarged. This was because the earth's surface is not flat. However, to be able to talk about, to draw, to refer to, and to think of real property parcels in two-dimensional terms by using maps, sketches, or plats is so necessary, basic, and instinctive that methods had to be developed that would be able to transfer the surface of the earth to a flat plane.

The systems that were developed are called "projection systems." There are many different versions of projection systems—too many to enumerate here—but the advantage of all projection systems is that, when applied, land surface areas can be described in two-dimensional terms. Straight lines can be parallel, and all straight lines will have constant bearings throughout. "Point," "line," "angle," and other geometric terms can now be correctly used in their two-dimensional sense to designate real property features. As long as the user is aware of the fact that a projection is involved, maps and property descriptions are readily developed and can, very accurately, reflect real property parcels. The few systems that are going to be discussed here are all "conformal" projection systems. This means that angles measured

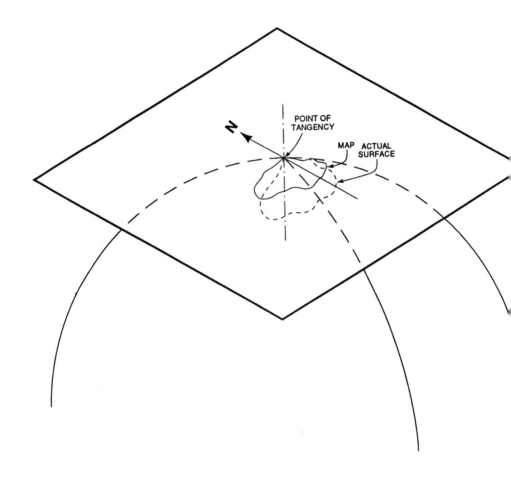

SKETCH 14

at the surface of the earth are presented on the projection (map) with minimum distortion.

5.2. TANGENT PLANE PROJECTIONS

The most common form of plane projection systems in use today is the tangent plane projection system. Although it is slowly being replaced by more sophisticated systems, most survey maps or plats are based upon this simple system. This is especially true of small surveys and surveys performed before the World War II. Modern, large-tract surveys are less likely to be based upon this system because of several disadvantages that affect accuracy and recoverability.

Sketch 14 is a visualization of a tangent plane projection. Imagine that a flat plane of glass is balanced on a globe. The plane would touch the globe in only one spot. This is the "point of tangency." If one were to look down through the glass from directly above the globe, features on the globe would be visible. These features could then be traced on the glass in their apparent positions as observed from above.

The tracings on the glass would be a map of those portions of the globe visible from above. The features mapped would become increasingly distorted as the distance from the point of tangency increased. The area very near the point of tangency would have minimum distortion and, depending on size and precision requirements, could become a very serviceable two-dimensional map of that portion of the globe.

The procedures for establishing a tangent plane projection in order to survey and map a parcel of land roughly imitate the preceding example. There are many variations of this procedure, some of which will be discussed later; yet all tangent plane systems have certain common factors. Concepts or steps that are underlined in this discussion will be those things common to all forms of tangent plane systems.

In order to *establish the point of tangency,* the surveyor may choose a location on or near the area to be mapped or surveyed. This location is the point where the surveyor *orients the work to some form of directional control* that is recoverable. By "recoverable" is meant physical marks on the ground, magnetic observations, celestial observations, or any procedure that defines a direction that can be repeated by another, at another time.

True north is widely claimed but almost never used as this control meridian. The most common form of directional control in boundary surveys of this type is an assumed north created by the recovery of a "reported bearing." A "reported bearing" simply indicates that two or more marks set by a previous surveyor were found and that the direction between these marks reported by the previous surveyor was used as the directional control for the present work. Sometimes the control direction may be entirely "assumed"; that is, the surveyor did not use a known or recoverable direction as a control.

The surveyor then *measures the angles formed between lines of the survey and the control direction.* (This is very different from independently measuring a magnetic or astronomic direction for each line, as is the case in projectionless maps.) The angles so measured during the survey may then be drawn, two-dimensionally and to scale, on the map or plat. More commonly, the relationship of the land lines in the survey with the control direction are expressed by reporting the bearings of the lines on the map. When such bearings are used, they are *two-dimensional computations based upon angles measured from the control direction.* Bearings shown on tangent plane projections, and even those correctly based upon true north, *do not represent the geodetic, astronomic, or true bearing of the land line,* except along the meridian that passes directly through the point of tangency. *The curvature of the earth and the convergence of the meridians are ignored.*

The distances measured during the survey are reduced to horizontal distances at a local datum. *No adjustments are made to compensate for differences between distances between points on this level surface and distances between the same points on the flat plane of the projection.* In short, the tangent plane projection system accounts for the curvature of the earth by *ignoring it.* Indeed, so little attention is given to the shape of the earth in performing tangent plane projection surveys that many people, even surveyors, forget that every map or plat derived from measured angles and distances is the projection of a curved surface onto a flat plane.

5.3. ADVANTAGES OF TANGENT PLANE SYSTEMS

The advantages of a tangent plane system are many and compelling. Chief among these advantages is the ease with which such a system is established. The surveyor has a wide choice of methods of directional control and complete freedom in choosing the point of tangency. This freedom reduces to a minimum the fieldwork required to map a particular parcel of land.

Knowledge of the elevation or geodetic position of the area being surveyed, in order to account for the earth's shape, is not necessary. Because the shape of the earth is ignored, there is no need to collect information on the geodetic positions of points within the parcel. This is a further reduction of the fieldwork required to survey and map a parcel of land.

The computations required are all two-dimensional and relatively simple. Field personnel do not need to master geometry beyond the high school level in order to understand and execute the data collection required. This reduces the formal training required of entry level survey crew members, as well as the on-the-job training needed to attain adequacy. Because training requirements are low, salaries paid to field personnel are also typically low, especially when the responsibilities and the value of the work performed are considered. Consequently, the initial or ''first-time'' costs of tangent plane projection surveys are comparatively low.

Straight land lines have constant bearings. Parallel straight land lines exist and have the same bearing. Landowners and other lay persons quickly accept and understand the maps or plats produced by this system, because the information conforms to their flat-earth instincts. The intricacies of defining land lines on the earth are greatly reduced and simplified.

Distances used in computations and shown on maps do not require a lot of "refining" and, in most cases, are simply a report of the level distance between land points at the average elevation of the site. Indeed, changes in elevation throughout the site are rarely great enough to be significant.

5.4. DISADVANTAGES OF TANGENT PLANE SYSTEMS

The disadvantages of tangent plane projection systems are as numerous, although not as obvious, as the advantages. Unless referenced to the same point of tangency, tangent plane surveys are like crackers floating in a bowl of soup: Each is free of the other without any ties or a common reference base. This means that a survey based upon the tangent plane projection will not contain any information about that particular survey's relationship to any other work that is not a part of the present survey.

Other surveys in the area, even those of adjacent properties, will disagree on the dimensions and directions of common lines. The directions between identical points on adjacent parcels will show different values. This lack of consistency of direction means that discrepancies, overlaps, disputes, or other irregularities will not be apparent by merely reviewing the survey maps of adjacent parcels. Potentially devastating conflicts in deeds may go unnoticed for generations until clear acts of possession take place.

Resurveys of real property parcels are entirely dependent upon recovery of survey marks or monuments set during the original work. This greatly increases the time required to retrace previous work. Failure to recover at least part of the original work will result in a retracement that has a poor probability, at best, of remarking boundaries in their original locations. Because permanence of boundaries is an integral part of the rights of possession, this is a most serious flaw.

Tangent plane projection surveys are of little use to land planners, municipal utilities, and other governmental agencies, because the lack of a common reference base makes the task of relating several parcels to each other very difficult. Multipurpose land records systems, so vital to proper community planning and resource planning, can make little use of the work of private surveys because of the lack of a common reference base.

Recovery of boundaries, resurveys, and resubdivision of existing parcels require extensive field recovery and frequent judgmental decisions by the surveyor. This greatly increases the costs of resurvey or recovery work and increases the possibility of conflicts. In the long term, tangent plane projection surveys can be much more expensive than more sophisticated systems.

This "hidden cost" limits the cost effectiveness of tangent plane projection surveys to very small, usually urban, parcels, where boundary lines are well established, clearly marked, and well maintained. Because the area surveyed under a tangent plane system must be kept quite small, or the factors ignored by the system will result in unacceptable irregularities, this eliminates the tangent plane system for use in large-tract surveys, highway routing, and state or county level property accounting.

5.5. STATE PLANE PROJECTIONS

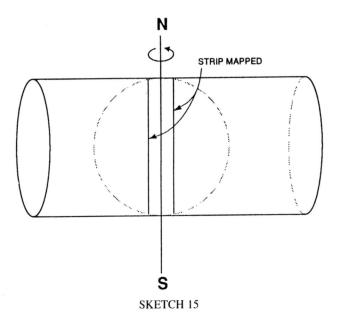

SKETCH 15

The rapid increases in land values, the need to mark real property boundaries permanently and precisely, the need for large-scale planning, and many other factors have led to the wide use of projection systems that eliminate some of the disadvantages of the tangent plane system. This need was so prevalent that, late in the nineteenth century, the federal government of the United States began promoting the use of particular projection systems adapted to each state. The particular type of state plane system that each state uses varies greatly, but the benefits, application, advantages, and use of these systems is the same for every state.

One form of state plane projection is demonstrated in Sketch 15. Instead of a rigid plane of glass, as in Sketch 14, imagine that a clear, flexible sheet of plastic is wrapped to form a cylinder around a globe. Then the sheet would touch the globe in a series of points instead of in one place. The features visible through the sheet could be traced onto the plastic, and, when unrolled, the sheet of plastic would form a flat plane on which a map of the globe would appear. The features on the map would become distorted as the distance to the line of contact was increased. The area near the line of contact would form a strip that would closely match the globe.

Let us further visualize that, instead of using a blank sheet of plastic, we begin with a clear sheet of plastic that has a rectangular grid pattern already established on it. This grid could consist of two sets of parallel lines intersecting at right angles, and each line would be defined by its distance from an established origin. The distance east on the grid is called the "x" distance. The distance north is called the "y" distance. When this grid sheet is wrapped around the globe, we could also deliberately align one of the grid lines with the north on the globe along the line of contact. Now, for each and every feature on the globe for which there exists a geodetic latitude and longitude measured in degrees, there can be found a corresponding grid location on the projection defined by the "x" distance and the "y" distance on the grid and measured in feet or meters, depending on the system.

Because the relationship of the sheet of plastic and the globe was rigorously controlled, then a specific set of such strips could be used to map the entire earth. In actual practice, this is what is done through mathematical manipulation of the geodetic positioning network. This procedure, and innovative variations of it, have produced mapping systems that cover large portions of every state such that most states are typically served by two or three overlapping projection zones. These state plane projection systems, called "state plane coordinates" (SPC), are defined for each particular state by that state's legislative body.

These systems are tied to the worldwide geodetic positioning system of latitudes and longitudes. The projection accounts for the convergence of the meridians, as well as the elevation of the parcel. The system mathematically and rigorously projects a portion of the earth surface onto a theoretical flat plane. The portion of the earth mapped is typically a strip, about 150 miles or so wide, wrapped around the earth. Most states are divided into overlapping zones in order to obtain full coverage.

The geodetic monumentation network that supports these projections is maintained by the federal agency known as the National Geodetic Survey (NGS). All national mapping is also controlled by this geodetic network. The control network includes both horizontal control (geodetic positions) and vertical control (elevations). Any survey work that is performed as part of this network is automatically tied to any other work in the system, be it private or public surveys.

The procedure for establishing a state plane projection survey requires that the surveyor recover one or more monuments of known geodetic position. The latitude and longitude of the known location is "projected" to the flat state projection plane by a mathematical process. The location of that point on the projection plane is defined on the two-dimensional grid. The "x" value defines the distance east, whereas the "y" value defines the distance north. The process is reversible: If the "x" and "y" of a particular location are known on any state plane projection system, the corresponding geodetic latitude and longitude of that location can be computed.

The surveyor then measures the angles formed between land lines and the distances between land points in much the same way as in the case of the tangent plane projection. The major exception is that the computations systematically account for the curvature of the earth by converting level distances to grid distances and by defining all directions in terms of grid bearings. The adjustments to bearings and distances are "hidden" and usually not presented on the finished plat. The final map or plat produced will show the level ground distances measured, not the grid distances used during computations.

5.6. ADVANTAGES OF STATE PLANE PROJECTIONS

The advantages of a state plane projection system are numerous, if not apparent. The primary advantage is that, because surveys performed under these systems become, in a sense, a part of the national geodetic network, the boundary corners defined in the survey can be related to all other surveys in the system. Boundaries, therefore, need not rely solely upon the recovery of marks within the survey in order for the surveyor to learn of that parcel's relationship to some other tract.

The recovery of boundary marks defined in state plane projections is greatly expedited by reducing considerably the area of search. This reduces the cost of resurveys, resubdivision, or recovery of existing parcels. Indeed, in areas where the use of state plane projection systems is widespread, even the costs of initial surveys is greatly reduced.

The information shown on adjacent tracts of land, and even remote tracts, can be related to the parcel surveyed. Surveys of adjacent properties will report (theoretically) the same bearings and distances for common lines. This makes the problem of discovering inconsistencies between adjacent tracts much simpler.

Municipal, county, and state level projects and planning are more easily controlled and regulated. Highways, pipelines, utilities, and other public works are defined in relationship to every private parcel surveyed under the state plane system. This enhances the acquisition of rights-of-way by increasing the precision of the description of the areas acquired.

The surveyor uses marks set by the NGS or other agencies that are at published locations and known to all. Marks that define horizontal location are called "stations." Marks that define vertical locations (elevations) are called "benchmarks." Both stations and benchmarks are part of an international network of control points and define locations based upon a particular datum.

In areas where the national control network is not well monumented, the state plane projection system can still be effectively used by a simple modification. If knowledge of the exact latitude and longitude of a point is not available (usually because of lack of monumentation), the approximate latitude and longitude, such as perhaps scaled from a U.S. quadrangle map,[1] are all that is required to convert astronomic directions in a particular area to precise grid directions. All the adjustments of the state plane system can be applied, and only the exact "x" and "y" of the survey points are excluded from the computations.

5.7. DISADVANTAGES OF THE STATE PLANE PROJECTION

The lack of control stations or monuments of known "x" and "y" in the immediate area of the survey may require extensive work beyond the area of the survey simply to make the "tie to the network" necessary for complete implementation. This disadvantage is being quickly eliminated by the introduction of artificial satellites in position determination. The geodetic position of locations on the earth can now be determined very precisely by using relatively inexpensive radio receiving stations in remote areas. The perfection of this method of determining location will mean that the type of control required by state plane projection systems will be available everywhere.

The computations and adjustments required in state plane projection systems are slightly more complicated than those required in the tangent plane system. The curvature of the earth and the convergence of the meridians are accounted for instead of ignored.

The educational and training requirements of field personnel performing

---

[1]These maps are printed by the U.S. government and are available, in various scales, covering the land mass of the United States and elsewhere. For more information, contact the National Cartographic Information Center, U.S. Geological Survey, 507 National Center, Reston, Virginia 22092, (703) 860-6045.

state plane projection surveys are more extensive. The field crews must possess or acquire knowledge of both the real property boundaries in the area of the survey and the geodetic positions of certain control points.

The additional training requirements and the additional fieldwork required often cause the ''first-time'' survey costs of state plane projection surveys to be more than those of tangent plane projection surveys of the same parcel. This is not true if there have been SPC-based surveys performed on adjacent tracts.

The few disadvantages cited indicate that, except for small urban parcels, the increase in costs for ''first-time'' boundary surveys are more than offset by the permanence, recoverability, and certainty of location that are rendered to real property boundaries documented by the state plane projection system.

CHAPTER 6

FUNDAMENTALS OF MEASUREMENTS

In the preceding chapters, we have made frequent reference to the *measurement* of angles and distances. It is very important to explain more fully the term "measurement." Theoretically, there is only one true distance between two land points at a particular level surface. Likewise, there is only one acute angle formed between two rays or lines. The act of determining the values associated with these angles and distances is the act of measurement. Unfortunately, even the finest of measurements is only an estimate. Indeed, *all measurements are estimates!*

Measuring is not the same as counting. Counting is dependent upon an indivisible fundamental unit, below which the thing or things being counted do not exist. The smallest unit of money is the penny. Even though we say that a dollar forms the basis of our monetary system, the penny is the smallest unit possible. Accountants do not measure the money in a bank; they count it. Empty your pockets and you either have 50 cents or you don't. Changes in quantity can only take place one penny at a time.

6.1. ACCURACY AND PRECISION

The terms "accuracy" and "precision" when referring to measured dimensions are not synonymous. "Accuracy" is an evaluation of the difference between a measured value and the "true" value. "Precision," on the other hand, is an evaluation of the procedure used to arrive at a particular value. For example, assume that the distance between two particular points is exactly 500 feet. Two men are requested to determine the distance by whatever method they see fit. Mr. Jones stands over one point and, peering at the other, announces, "500 feet." Mr. Smith acquires a 100-foot measuring tape and carefully measures the distance several times and declares it to be 500.02 feet. Purely by luck, Mr. Jones's estimate was more accurate. Mr. Smith's estimate, because of procedure, was more precise.

6.2. IMPLIED PRECISION

Measurement data are usually presented by parameter (feet, miles, degrees, etc.) and numerical values. The form that the data are presented in contains a clue to the precision of the measurement procedure used to arrive at the value. This is only a clue, for unsophisticated reporters of measurement information can unwittingly change the implied precision of a reported measurement.

The nature of the parameter is important when implied precision is analyzed. A distance given in paces implies a precision far below that of a distance given in inches. A direction given in "points" of the compass (i.e., northeast) implies a precision below that of a direction given in degrees.

The division of the parameter is also significant, for it contains information about the smallest unit of measure used. A dimension reported as 22 feet implies that a foot was the smallest unit of measure. A dimension reported as 22 feet, 0 inches implies that an inch was the smallest unit of measure. The last figure given is said to be the "doubtful" figure in a measurement process. If a measurement were reported only to the nearest foot, then distances from just over 21 feet, 6 inches to just under 22 feet, 6 inches would all be reported as 22 feet. If measurements were reported to the nearest inch, then distances from just over 21 feet, $11\frac{1}{2}$ to just under 22 feet, $0\frac{1}{2}$ inches would be reported as 22 feet, 0 inches. The implied precision concept classifies 22 feet and 22 feet, 0 inches as two different measurements!

Fractions form another category under the implied precision rules. "$\frac{1}{2}$ foot" is not the same as "6 inches," nor is either one the same as "0.50 feet." The implied precision of $\frac{1}{2}$ foot is that the smallest unit of measure was $\frac{1}{2}$ of a foot. The implied precision of 6 inches is that the smallest unit of measure was an inch. The implied precision of 0.50 feet is that the smallest unit of measure was one-hundredth of a foot. $\frac{1}{4}$ of a degree is not the same as 15 minutes, and 1 acre is not the same as 43,560 square feet when we are considering implied precision! Many a realtor has multiplied an area given in acres by 43,560 (square feet in an acre) to arrive at a square footage in order to determine a sale value. If a area is given in even acres, the square footage, under the implied precision rule, could be off by 21,780 square feet either way![1]

6.3. ERRORS

Measuring is typically dependent upon the comparison of the thing being measured to a standard or known value. Variations in the accuracy of the standard value, the precision of the limits, as well as a host of other factors, introduce variables into the measurement process. These variables are called "errors." The word "error" has a very different meaning to the surveyor or scientist than it does to the lay person.

There are three general categories of errors: systematic, random, and blunders.

6.3.1. Systematic Errors

Systematic errors are inaccuracies in measurement that occur in the same direction every time that a measurement takes place. If, for example, a ruler were a fraction of an inch shorter than the length shown on it, then every measurement using that ruler would be off by the same amount and would mislead the user in the same direction. Objects would consistently be reported as longer than they really were.

6.3.2. Random Errors

Random errors are inaccuracies in measurement that occur in varying directions and magnitudes. Every time a carpenter measures a cut line, he or she must estimate the alignment of the end of the ruler with the end of the wood and the alignment of the marking pencil with the desired measurement. Each time these estimates are made, the carpenter might be off to the right one time and off to the left another. Random errors tend to cancel one another out when measurements are repeated several times.

6.3.3. Blunders

Blunders are not really errors in the strict scientific interpretation of the word. A blunder is the gross misinterpretation of a measurement due to a

[1] For a more detailed analysis on the theory of measurement, the reader is referred to R. B. Buckner, *Surveying Measurements and Their Analysis* (Rancho Cordova, CA: Landmark Enterprises, 1983) and E. M. Mikhail and G. Gracie, *Analysis and Adjustment of Survey Measurements* (New York: Van Nostrand Reinhold, 1981).

careless or mistaken observation. The carpenter meant to cut a 4-foot plank but marked the cut line at 3 feet. "Measure twice, cut once" is the old saying that addresses this type of blunder.

6.4. ALL MEASUREMENTS INCLUDE ERRORS

All measurements include errors. Imagine the following experiment. A classroom of high school students was instructed to attempt to measure the teacher's desk using a feed store yardstick. The teacher requested that each student measure each dimension to the smallest fraction of an inch that he or she could estimate. As one would reasonably expect, some variations in the reported dimensions of the desk were reported. The values for the width, for example, might have ranged from 40 1/32 to 40 7/32 inches. The average width reported would then have been 40 $\frac{1}{8}$ inches. One might report that the width of the desk was 40 $\frac{1}{8}$ inches, with a measurement error (uncertainty) of plus or minus 3/32 of an inch.

The true width of the desk is not known and never will be known. The absolute, exact distance from the first molecule of desk on one end to the last molecule of desk at the other end is much finer than 1/32 of an inch. The feed store yardstick is not a calibrated standard, and changes in humidity and temperature affect the length of the yardstick and the width of the desk. The variety of skill possessed by the many different measurers also influenced the results. The difference between the absolute width of the desk and the reported average width is the measurement error.

The students were engaged in a direct measurement. The standard (the yardstick) was applied directly to the desk. The beginning and end points were the limits of the value sought. This is the simplest act of measuring a quantity. If the students wished to reach a more refined value, they might have recorded humidity, pressure, temperature, and other factors and examined how variations in these factors affected the length of the yardstick. The students might have compared the length of the yardstick with a calibrated standard so that corrections for imperfections in the yardstick could be accounted for.

The problem of estimating the fractions of an inch might have been simplified by the addition of a vernier[2] or some other device that reduces the amount of interpretation involved. These are things that increase the precision of the measurement of the desk, but the fact remains that the measurement will never be absolute. *No matter how refined the process is, there*

---

[2]An auxiliary scale, mounted on the primary scale or ruler, that permits a direct reading of fractions (usually tenths) between divisions marked on the primary scale.

will remain a range of values that can result from correct and legitimate application of any measurement procedure.

6.5. REDUCTION OF ERRORS

The science of measurement is the study of methods or procedures that eliminate blunders, account for systematic errors, and reduce the effect of random errors. Some of the methods by which blunders can be eliminated are by repeating measurements, by careful observation, or by analysis of the results. Some of the ways in which systematic errors can be accounted for include standardization of the equipment used, understanding the effects of the environment, and analysis of the results. Systematic errors must be detectable before methods to account for them are possible. Some of the ways by which the effects of random errors are reduced include refinement of instrumentation (less estimation in interpreting readings), multiple observations, and analysis of results. How this is done in the modern land survey will be discussed later.

6.6. DEVELOPMENT OF STANDARD PROCEDURES

The history of land measurements is as ancient as the human race. The development of methods of identifying locations discussed earlier were accompanied by attempts to perfect the measurement of angles and distances. Ancient societies that developed a centralized government soon employed people whose task it was to measure the area under the control of the central government. The earliest reports of these specialists come from Babylon, where pacing was extensively used to measure long distances. These professional pacers would report the distances between towns so that the king might know how many days it would take for his army to travel from place to place.

The Egyptians, long noted as master builders, used ropes of known length that were knotted at regular intervals in order to measure the fields and to lay out the great construction projects for which they are so well known. Many of the words we use today originate from the methods used to measure the land. The word "mile" is derived from the Latin "milia," meaning "a thousand." The Roman soldiers would count each time the right foot struck the ground while marching. Each thousand paces was a milia. The average man today will cover about 5 feet for each pace (two steps), or 5,000 feet for a "milia."

Whether counting paces, turns of a wheel, camel paces, or any of a number of innovative means of measuring distances, the problem of a lack of a standard and inconsistency of results was a constant problem. The development of the Gunter's chain, by Edmund Gunter circa 1620, was the first real advancement in the measurement of horizontal land distances. The concept of standard length, horizontal measure, correction for temperature, and sag quickly followed this development. The Gunter's chain remained state of the art for measuring horizontal distances in land boundary surveys until the twentieth century.

The modern surveyor's steel measuring tape, still called a "chain," only improved slightly the measuring process and accuracy. It was not until the 1960s that the advent of the electronic distance measuring instrument significantly improved the accuracy of distance measurement. Even with these new and fantastically accurate devices, there still exists a measurement error that cannot be eliminated. It is for this reason that many states have developed a set of standard survey criteria listing the size of error acceptable for various grades of survey work. Table 3, in the Appendix of Tables, is an example of a typical set of measurement standards.

Unlike distances, the measurement of angles developed to a high precision very early in history. The capabilities of the modern surveyor to measure angles has improved only slightly during the latter part of this century. Although the country surveyor of years ago probably did not avail himself of the most sophisticated angular measurement devices possible, the capability was there. Indeed, angles were rarely measured to any great precision until the art of measuring distances improved to the point where real precision was possible.

Yet, even when calibrated metal tapes are applied directly between boundary markers and all of the appropriate corrections are made, there is a range of values that will be observed. The conditions that affect this range are sometimes beyond the control of the observer. Size, shape, and the nature and condition of the end points of the land line segment being measured is a significant factor.

If the northeast corner of farmer Jones's property is marked by a 6-inch square concrete post, 4 feet of which is above the ground and leaning southwest several inches, while the northwest corner is marked by a 4-inch diameter post, also several inches out of plumb, then even the most carefully applied measurement procedures are not going to be more precise than plus or minus a few inches. Before beginning the measurement, the surveyor has to estimate the upright location of each corner and the center point of that upright location. The condition of the corner monuments are not such that a more precise measurement can be made.

The measurement of distances is not the only kind of measurement required in real property surveys. The angles formed at the corners of property parcels are also measured. The condition of the end points, uncertainty of location, and limitations of the instruments involved mean that the angles measured, like the distances, have inherent measurement errors that must be understood and accounted for.

Boundary lines are, more often than not, subject to acts of possession or acquiescence. Fences, hedges, tree lines, or other physical obstructions frequently occupy the full length of a real property boundary. These features, while clearly indicating the general location of the boundary, make the direct measurement of the distance between corners quite difficult. The vast majority of boundary dimensions in the world are the result of indirect measurement. Dimensions are almost always the result of computations based upon several measurements of angles and distances. In light of this, it is a constant source of wonder that so many real property boundary dimensions are the same from survey to survey.

6.7. COLONIAL PERIOD

Perhaps the best way to clarify the impact of measurement methods, procedure, accuracy, and reliability upon modern real property parcels is by reviewing how the procedure and precision of the past compare with those today. For the purpose of illustration, let us imagine that a particular boundary in colonial America had each end of the line marked by 6-inch square stone monuments. The royal surveyor in 1770 may have been required to produce a survey plat (Sketch 16) to assist in the identification of the parcel to be transferred to private hands.

The royal surveyor probably would have used the following method to determine the distance between the corner markers. Because the line in question is a boundary line, there would probably be a fence or a hedgerow along the entire length of the line. The royal surveyor would, therefore, clear a line parallel with the boundary at some convenient distance from the actual line. Temporary wooden pegs might have been set at each end of the cleared line opposite the stone corner markers. These pegs would have marked the ends of the segment to be directly measured.

Although a Gunter's chain might have been available, a device called a "compass" or a "toise" might have been used to measure the distance between the pegs of the offset line. Sketch 17 provides a sample of how a toise might have been constructed. Depending upon where in colonial America the land was located, the dimensions of the toise would have varied from about 6 to 6.4 feet. The device would usually have been made of wood by

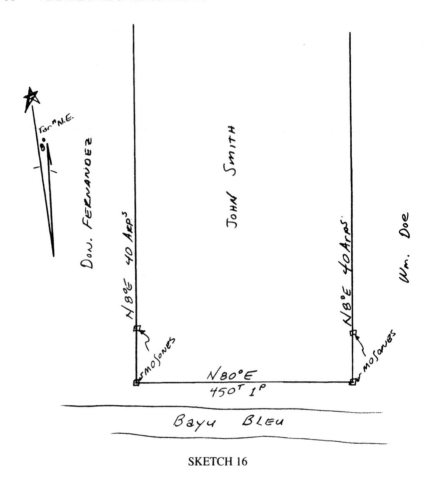

SKETCH 16

the surveyor himself. The points might have been metal tipped or simply sharpened wood.[3]

The surveyor would have begun the measurement by placing one end of the toise at a peg and the other end of the toise on the ground in line with the far peg. The device would then have been rotated about the leading tip so that the trailing tip would be brought to the leading position and in line with the far peg. The process would be repeated, alternating the tips of the toise until the far peg was less than one rotation from the leading tip. This remaining portion would be measured with a yardstick or a foot rule.

An examination of this process reveals several sources of measurement

[3]In other areas, a pole or a rod 16 ½ feet long and possibly capped with brass ends was commonly used to measure distances.

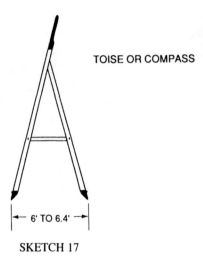

TOISE OR COMPASS

|← 6' TO 6.4' →|

SKETCH 17

error or inconsistencies. Any slight failure of the toise to be exactly the correct length would be repeated for every measurement. If, for instance, the distance between the tips were an eighth of an inch shorter than the 6 feet it was believed to be, then the surveyor would have reported a mile (5,280 feet) measured when only 5,270 feet, 10 inches had actually been traversed.

Each rotation of the toise would have to have been without any slippage. Any movement of the tips during the act of transferring the trailing tip to the leading position would introduce error. The work would often be done from horseback, adding another factor to the difficulty involved in simply rotating the device. Each tip would also have to be set perfectly in line with the end pegs, or additional error would be introduced.

The toise would have been set directly against the surface of the earth, so every rise and fall in the land would have added apparent distance to the measurement. The surveyor would sometimes have attempted to account for this increase in distance by estimating the amount of extra turns the changes in elevation caused. The practice of "adding one for good measure," that is, deliberately adding an unreported turn on the toise or adding one more length of rod to prevent "shorting" the buyer, was very common, especially in hilly country.

The beginning peg and ending peg would have been set in locations estimated to be directly across from the end points. This estimation would introduce error that could be quite large if the offset line were some distance from the boundary line. The surveyor would have recognized that the measurements made by him were far from perfect, and so the slight error introduced by estimating the offset peg locations would be considered acceptable. In cases of original surveys, meaning surveys that created new parcels, the lines run might have been the actual boundaries.

In measuring 1 mile, the surveyor would have to have made 880 rotations of a 6-foot toise. It would have been highly possible, even probable, that occasionally a miscount might have occurred. Often lines were measured several times to avoid miscount blunders, but just as often they were not. This factor alone introduced considerable doubt about just how much reliance could be placed upon surveys conducted in this manner.

The direction of the boundary line would be determined by placing a magnetic compass at one offset peg and sighting the other. The bearing of each boundary line would be determined in the same way. The angles formed at the corners would not be measured in most cases.

The preceding scenario assumes that the end points or offset end points were intervisible. If the intervention of hills or woods made it impossible to see one corner post from the other, the problem of aligning the route of the offset line would be much more complicated. In this case, a magnetic compass would be used to maintain line as well as to determine direction.

If the end points were not intervisible, the surveyor would set a magnetic compass over one offset peg and send a rodman as far ahead on the line as could be seen, using the compass bearing believed correct for that boundary. The distance to the rodman would be measured as before, and the compass would be brought ahead to the rodman. The rodman would then be sent ahead again and the process repeated until the end peg was reached. The distance from the offset line at the end of the segment being measured would be noted and corrections to the reported bearing would be made if the end offset was significantly different than the starting offset.

If the work being done was the creation of a new parcel, then the actual boundaries would usually be traversed in a consecutive sequence such that the surveyor would begin the work at one corner (the point-of-beginning) and measure the bearing and distance of each side, in order, returning to the point-of-beginning. If the new parcel was to be *a specific size* then one additional step would be added.

Because the new parcel being created represented a bounded area, the surveyor would know that, if the measurements were perfect, then the last line measured would return to the point-of-beginning. Of course, perfect measurements are impossible, so the last line measured would never return to the exact point-of-beginning. The most common method used to address this problem would have been the "field correction."

After the surveyor had satisfied himself that some gross blunder had not occurred, the "field correction" would be applied. This correction usually would consist of adjusting the location of one or more of the survey marks

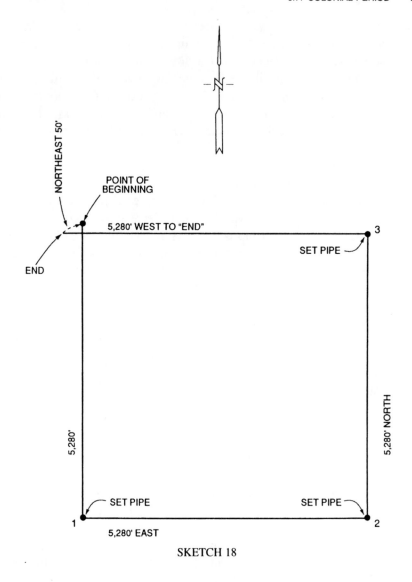

SKETCH 18

until the *distances* called for were satisfied. In Sketch 18, the attempt to return to the point-of-beginning resulted in arriving at the point labeled "end." In this case, the surveyor might have applied the field correction by simply moving the point labeled "3" northeast 25 feet. Each surveyor had his own preferred method of applying field corrections, and, in most cases, these corrections were applied based upon the surveyor's judgment, not upon any scientific analysis.

The procedures outlined were representative of the type of work that was done in the latter part of the eighteenth century. The actual practice in any particular part of the country varied greatly. The notes, letters, and drawings of the surveyors of that period in an certain area would be the best source of information on actual practice in that area. Plats often hold clues to the procedure used. Because of the uncertainties involved, the distances reported by these early surveyors could not be expected to be more accurate than 1 foot in one hundred. The bearings shown are even less reliable because of the lack of precise sighting devices and reliance upon magnetic north.

The concepts of errors in measured values were well known in the eighteenth century. The degree to which the methods of accounting for these errors were applied is a reflection of land values, craftsmanship, and skill or sophistication. Multiple observations might have been used to eliminate blunders. Skilled craftsmanship could have reduced the magnitude of systematic errors. But the greatest tool in the detection of errors, the analysis of results, was not very effective in the eighteenth century, because the measurements of direction were so crude compared with the measurements of distances.

If one were to measure a closed area, such as a parcel of land, a mathematical analysis of the distances and directions reported could determine if the values presented were consistent with each other. During the eighteenth century, the crudeness of the directions observed, the laboriousness of the computations, and the low value of the land meant that these methods were rarely applied.

6.8. POST–CIVIL WAR PERIOD

If the imaginary boundary shown in Sketch 16 were resurveyed a century later, there would have been only a few changes in the survey procedure. The existence of a fence line or a tree line along the boundary would have still required the use of the offset survey line. Pegs would have been set at estimated right angles to the boundary line a short distance from the stones just as 100 years ago.

The surveyor of 1870 probably would have used a variation of the Gunter's chain, called the "two-pole chain," to measure the distance between the end pegs. The Gunter's chain was a iron chain of 100 links that was 66 feet long. Sketch 19 is a detail of a portion of a Gunter's chain. The sketch is based upon an antique that had tags on each link, which was not a common practice. Link tags were more commonly set at 10-link intervals or more. The two-pole chain was a shortened version of the Gunter's chain and was 33 feet long and had 50 links. A two-pole chain was laid out twice to measure one "chain" of distance.

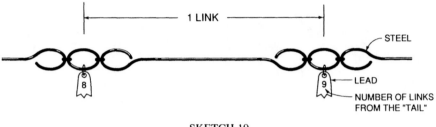

SKETCH 19

The measurement of the distance began in the same way as the 1770 survey. One end of the chain was held over the starting peg, and the chain was stretched ahead, supported at the ends only. If the ground was sloping, the ends were held such that the ends of the chain would be at roughly the same elevation. This was done so that only the horizontal distance was measured. The end of the chain was marked by a pin, and the chain was moved forward. Alignment was maintained by sighting the end points, if possible, or by sighting the bearing of the line with a magnetic compass.

The common sources of errors in the process were well known, and some attempt at compensating for these errors was often made. The chain length was subject to wear, because the chain links rubbed together. Often a standard chain was kept at the home office for comparison with the field chains. The sag of the chain supported at the ends varied with the force exerted by the chainmen. The marking of the chain ends by pins sometimes included guesswork as to where the ends actually were, especially in rough terrain. Failure to measure in a straight line also introduced error.

The sighting devices on the surveyor's compasses were improved, and the introduction of the solar compass slightly reduced some of the uncertainty associated with magnetic directions. Generally speaking, the state of the art had advanced to the point where distances measured were accurate to 1 foot in 500. Directions were probably accurate to 1 degree, although failure to account correctly for magnetic deviations still rendered this part of the surveyor's work the least accurate.

The slight improvement in determining directions also improved the effectiveness of analysis of results in detecting errors. Although mathematical analysis was possible, time, cost, and the specter of laborious computations meant that any analysis of results usually took the form of plotting the parcel to scale to determine if it "closed." The use of mathematical analysis did begin to increase with the approach of the twentieth century. This is not to say that rigorous mathematical analysis of survey results did not take place prior to the twentieth century; only that it was not commonly used. Sketch 20 shows how that same parcel might have been platted in 1870.

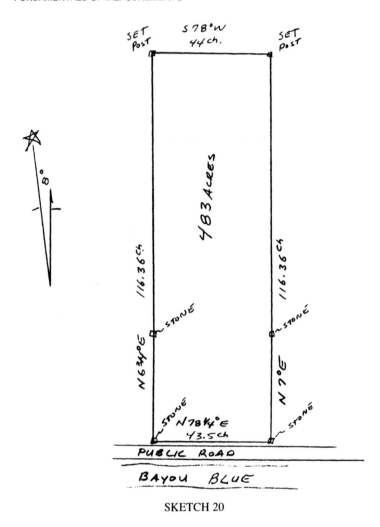

SKETCH 20

6.9. BEGINNING OF THE MODERN PERIOD

By 1920, the introduction of the surveyor's transit[4] into everyday survey work had corrected some of the shortcomings that had hounded surveyors in the past. Instead of observing independent bearings for every line of a boundary, the surveyor now measured the *angles formed at the boundary corners*. This greatly increased the reliability of the reported shape of the parcel of land being surveyed, which greatly increased the accuracy of the

---

[4]In America, this term is used to identify a theodolite that is read by use of a vernier scale and has a telescope that can be inverted (transited) without removing it from the instrument.

acreage computed, but it did not improve the reliability of the magnetic direction used to orient the survey.

Distance measurement was much the same as in 1870, except that a steel surveyor's tape (still called a "chain") measured the distance in feet and decimals of a foot. The actual distance measured was still between offset points. The relationship of the offset points to the corner monumentation was now moie accurately known because of the use of the transit to measure the angle to the actual corner. The steel tape was laid out and the lengths marked with chaining pins exactly as the two-pole chain had been used. Sketch 21 shows how our imaginary parcel might have been platted in 1920.

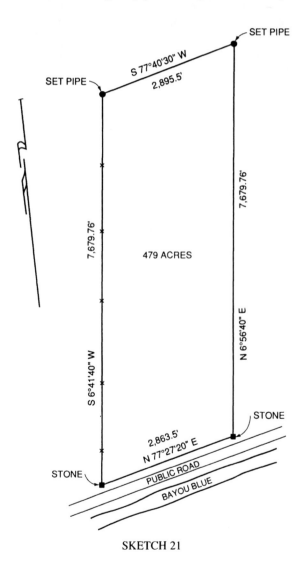

SKETCH 21

The introduction of the transit enabled the surveyor to measure accurately the angles formed by the intersecting boundaries of the parcel instead of simply measuring the independent bearings of each side. This increased the accuracy of the typical survey of the day to the point where deviations of less than 1 foot in 5,000 were common. For most work of the early twentieth century though, the accuracy of 1 foot in 1,000 for distances and 1 minute of arc for angles is about the norm.

The introduction of the transit to measure angles also meant that the mathematical analysis of the distances and angles measured became a very useful tool in determining the validity of the results of a survey. The concept of "closing a traverse" became widespread. If a parcel of land were measured by linking all of the lines around the parcel, then a surveyor might compute the changes in distance and direction that took place if he or she were to travel from one corner to another in consecutive order. This "traverse" would return the surveyor to the exact point-of-beginning if all of the angles and distances were error free (assuming a proper plane projection was used). The amount by which this traverse failed to return to the point-of-beginning would then be the sum of all the measurement errors: systematic, random, and blunder.

If a blunder occurred, then a large failure to close would have been noted (unless two blunders in opposite directions occurred, in which case one might hide the other). If the failure to close was within an acceptable range, then the surveyor would "balance" the computations by simply distributing a "correction" to each angle and distance measured until the figure did "close" upon mathematical analysis. Variations of these methods of "balancing the traverse" are used today but with greater sophistication (we hope).

The most common methods of "balancing a traverse" generally available were, and still are, the compass rule, the transit rule, Crandall's rule, and the least squares adjustment. Notice that, although the first three methods are identified as rules, the last is identified as an adjustment. All of these methods, with the exception of the least squares adjustment, require the use of a plane projection system, where each point can be defined by plane coordinates. Least squares adjustments can be applied to any set or group of measurements, regardless of the mapping or computational control used.

6.9.1. Compass Rule

The compass rule was, by far, the most popular method of balancing a traverse during the early and mid-twentieth century. It is still quite popular today and, under certain limited circumstances, is still a valid method.

In Sketch 22, the measurements made by the surveyor on the example

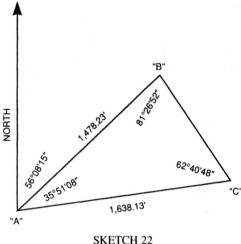

SKETCH 22

triangular traverse are shown. It is known that the interior angles of a closed figure must add up to a particular value; in this case, 180 degrees. The sum of the measured values is 179 degrees, 58 minutes, 48 seconds, indicating an error of 1 minute, 12 seconds. The first step in applying the compass rule to this case would be to divide the total angular error by 3 and add this number to each of the three measured values. This adjustment of angular error is made *before* any other computations are made. Sketch 23 shows the "adjusted" angles of the example traverse. Note that no correction is made to the angle from "north" to the line "A–B."

Using the adjusted angles, a plane projection bearing of each line is com-

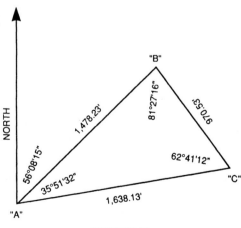

SKETCH 23

puted. These bearings are then used to compute the location of each of the corners based upon the distance north (or south) from the point-of-beginning. This distance is sometimes called "latitude" when it is referring to the change from one point to the next. Similarly, the location of each corner based upon the distance east (or west) is computed. This distance is sometimes called the "departure" when it is referring to the change from one point to the next. All computations are two-dimensional and are made on the theoretical plane of the map projection, even though the three-dimensional terms of "latitude," "longitude," "north," and "east," among others, are frequently used.

Consecutive computations of each point in the traverse will result in two sets of values for the point-of-beginning: the first set being the values assigned at the beginning of the computations and the second set being the values computed at the end. The difference between the original values and the computed values, or failure to close, has a north–south component and an east–west component. The compass rule states that a correction will be added to (or subtracted from) the computed coordinates of each point in proportion to the accumulated distance from the beginning, measured along each side, divided by the total distance around the traverse.

If, in the example traverse, point "A" were given the initial value of 0 feet north and 0 feet east, then the following table gives the initial computed values for the points shown, rounded off to the nearest one hundredth of a foot. A negative value in the "Feet North" column indicates south; in the "Feet East" column, it indicates west.

Point	Feet North	Feet East
"A"	0.00	0.00
"B"	823.67	1,227.49
"C"	−53.65	1,642.51
"A"	3.43	5.38

In order to "correct" the value for point "C," it is necessary to subtract 2.06 from the north coordinate [$3.43 \times (1,478.23 + 970.53)/(1,478.23 + 970.53 + 1,638.13)$] and subtract 3.22 from the east coordinate [$5.38 \times (1,478.23 + 970.53)/(1,478.23 + 970.53 + 1,638.13)$]. The coordinate for "C" can be "adjusted" to north −55.71 feet and east 1,639.29 feet. Each point in the traverse is corrected in this manner.

Once the coordinates are adjusted, the distances and directions between each point can be *computed* based upon the adjusted coordinate values. Sketch 24 shows the example traverse after an application of the compass rule.

The *adjusted values* are used in all reports, other measurements, loca-

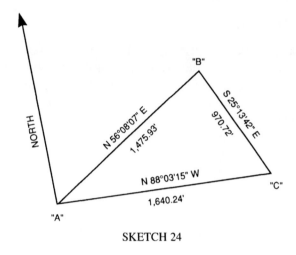

SKETCH 24

tions, and descriptions of the parcel being surveyed. Unlike the field correction, *no points are moved.* The only change that takes place is in the values assigned to the directions and distances of the lines.

One of the most common mistakes made in applying the compass rule is the failure to adjust the measured angles *before* the initial computation bearings and coordinates. This blunder, while resulting in an adjusted traverse, yields a different adjustment for the same traverse computed clockwise than counterclockwise. Another common misconception is that the compass rule adjusts the traverse based upon the probability that certain measurements will have greater error involved than others. In truth, the compass rule, like all the "rules," only *"makes things fit."*

6.9.2. Transit Rule

The transit rule also begins with an adjustment of the measured angles in the exact manner as the compass rule. Just as in the compass rule, the initial bearings of the sides of the traverse are computed from these adjusted angles. The coordinates of each point are computed using the measured distances and the adjusted bearings. The difference between the compass rule and the transit rule is in how the failure to close is proportioned to each coordinate.

The transit rule uses the ratio of the changes in north (or east to correct the east coordinate) for a particular line to the arithmetic total of the changes in north (or east for the east coordinate) and adjusts each latitude (or departure) of each line accordingly. This adjustment method was quickly recognized as one that produces inconsistent results, and so the use of this method has been, for the most part, discontinued.

6.9.3. Crandall's Rule

Crandall's rule was an attempt to apply the concepts of probability to error correction. According to this rule, errors in long distances are more probable than in short distances, and errors in distances in general are more probable than errors in angles. The application of this rule is a fairly complex computational procedure that essentially ignores angular errors and disproportionately adjusts distances. Crandall's rule almost always results in adjustments that are unnecessarily distorted, so the use of this procedure has generally been restricted to academicians.

6.9.4. Least Squares Adjustment

The least squares adjustment of a traverse is one of the many applications of the theory of least squares analysis of measured values. Unlike the rules cited earlier, the least squares adjustment is strictly based upon the theory of the propagation of measurement errors. The theory states that, for any set of measured values, the best set of corrections to apply to the measured values is one such that the sum of the *squares* of all of the corrections is minimized. The least squares adjustment is the most commonly used method of adjusting measured values.

The simplest example of the least squares adjustment theory is the average. If the least squares theory is applied to a single set of things that are measured many times, calculus renders the arithmetic average as the least squares solution. If the distance between two points were measured many times, then the average (obtained by summing all of the measurements and dividing by the number of measurements made) would represent the most probable measured distance. Every person has, at one time or another, used a least squares adjustment without knowing it.

The least squares adjustment is applied directly to measured values and, in surveying, is simplest when the precision of angular measurements is comparable to the precision of the distance measurements. If the procedure used to measure the angles (distances) is much more precise than the procedure used to measure the distances (angles), then special steps within the least squares adjustment must be taken for the results to be valid.

Unlike the rules, the angles of the traverse are not adjusted prior to beginning the least squares procedure. Angles and distances are adjusted simultaneously, based upon the theories of probability. The procedure renders consistent and reliable results in proportion to the quality of the measurements made.

6.10. MODERN PERIOD

From the mid-1970s to today, the increased use of electronic distance measuring devices (EDMs) has greatly increased the accuracy of the distances reported on modern surveys. Unlike the toise or the chain, the EDM does not rely upon repetitively "laying out" a standard length. Most EDMs measure distances by emitting a laser light that is reflected back to the instrument by a special mirror. The distance to the reflector is determined by comparing the departing signal with the returning one. The accumulation of error with every length of chain, so inherent with the old methods of measurement, is almost absent with the EDM. It is only necessary to be able to see from one end of the line segment being measured to the other in order to measure the distance.

The high precision of the EDM has matched the precision long possible in angular measurement. This "matching of precision" is perfectly suited to the least squares adjustment method of analyzing measured values. When the distance values were obtained by chaining, even when strict procedures were used, the angular values were usually more reliable. The surveyor had to determine the relative degree of difference in reliability in order to use the least squares adjustment method properly. This difficulty, which discouraged the use of least squares, was eliminated by the EDM

Distances of several thousand feet are measured as quickly and as easily as a few feet. This has also led to a more widespread acceptance and use of the state plane projection system by the professional surveyor. Full use of state plane system requires that the work be tied into control stations that may be miles from the job site. Before the accuracy and ease of long-distance measurement were provided by the EDM, one could hardly fault the surveyor in private practice for opting not to spend several extra days on a job just to tie it to the state plane system.

The introduction of the computer has allowed the surveyor to use formerly very cumbersome, but theoretically superior, computational methods of detecting error distribution and balancing traverses. Now, instead of using guesswork in identifying and correcting for errors, complex and sophisticated mathematical procedures are available to every surveyor. The formerly time-consuming tasks of computing traverses, areas, distances, and a thousand other things are now performed in fractions of a second. This relief from tedium has allowed the surveyor the opportunity to look at every problem from many different sides.

Modern surveys are performed with a degree of precision that was nearly impossible just a few decades ago. Accuracies of 1 in 70,000 are now commonplace. Angles and distances can be measured with equivalent degrees of accuracy. This results in a greater consistency or reliability of computed dimensions. This is perhaps best illustrated by the standard deviations for typical modern instrumentation listed in Table 5 in the Appendix of Tables.

6.11. RANDOM TRAVERSE

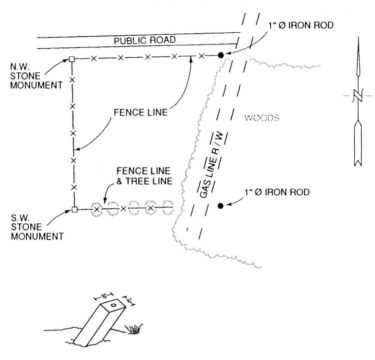

DETAIL OF S.W. STONE MONUMENT

SKETCH 25

In order to illustrate better the impact of these modern advances on the profession of surveying, let us recreate a typical boundary survey of a simple fictitious parcel of land. Sketch 25 shows the corner monuments of the parcel in their relative positions. The methods used to recover these corners will be discussed in detail in Chapter 8. The corners are the limits of the parcel, and the task at hand is to *measure* the distances between the corners, the angles formed, and the area of the parcel. Knowing how this is typically done today will greatly enhance your understanding of boundary surveys.

Because the property involved is valuable and because of the expected use of the property, the surveyor, in our example, has decided to use the local state plane coordinate system to control and map the survey. The sur-

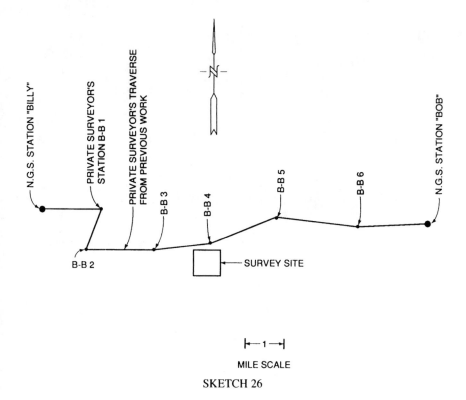

MILE SCALE

SKETCH 26

veyor may have previously established a "control traverse" between two NGS stations, "Billy" and "Bob," as a part of another job. The latitude and longitude of "Billy" and "Bob" are published by NGS and are the results of measurements that have been *adjusted to conform to* the national network. The latitude and longitude of these stations have been mathematically translated to the appropriate state plane grid coordinates ("x"s and "y"s), called the "values" of the stations.

The prior work was one of high precision and was rigorously adjusted to the national network. This means that the measurements of the traverse that were used to develop the "x" and "y" values for each traverse station were strictly controlled and that these values were *adjusted* to conform to the published values of "Billy'.' and "Bob." Consequently, the distances and angles actually measured in the traverse are not exactly the same as the distances and angles indicated by the assigned values of the stations. This variation between "measured" and "reported" is very small because of the high precision involved. Sketch 26 shows the relationship of this traverse to the parcel being surveyed. The traverse stations are often called "random," because they do not represent real property boundary corners.

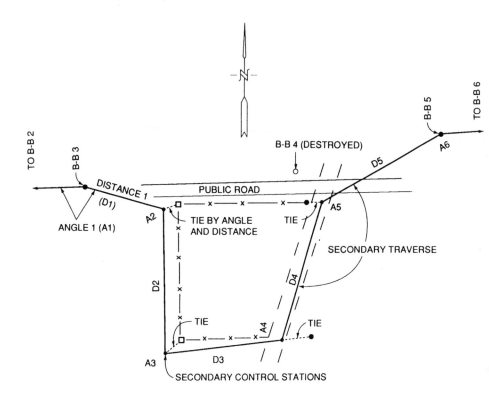

SKETCH 27

The land between the corner monuments consists of hedgerows, fence lines, heavy woods, hilly terrain, and other obstacles that prevent direct measurement between the corners. Therefore, the surveyor must establish a secondary control traverse that takes advantage of roads, power lines, pipelines, cleared fields, or other features that allow for unobstructed lines of sight. The points at which the secondary traverse changes direction are typically called "control stations" or "random stations." The locations of these control stations are deliberately chosen so that features that are to be recorded are near or, at least, visible from one or more control stations. Sketch 27 shows the secondary control traverse for this example.

Angles 1 through 6 were measured using procedures and equipment that resulted in quantities that were expected to vary from the actual value less than 5 seconds either way. Distances 1 through 5 were measured using procedures and equipment that resulted in quantities that were accurate to within 1 in 70,000. The values of each of the secondary stations is then computed and adjusted using the least squares method of traverse adjustment to conform to the values of stations B-B 4 and B-B 5. Once again, the adjustment must be within certain limits to be consistent with the level of precision required.

Each of the corner monuments is then tied into the secondary traverse by measuring the angle and distance from the secondary stations to the corner monuments. These ties must be multiple and redundant to avoid the possibility of blunder. "Multiple" means that the angles and distances must be measured several times. "Redundant" means that separate sets of measurements must be made wherein any one of the sets is sufficient to compute a value for the corner. The coordinate value for each corner is computed separately for each set of measurements, and the results are compared for consistency. The results are usually combined and averaged to produce the accepted coordinates for each of the corner monuments.

Once the coordinates for the corner monuments have been determined, it is a simple matter to compute the bearings and distances between each of the corners. The area of the parcel is also computed directly from the coordinates. The state plane projection requirements for accounting for the curvature of the earth and the conversion to and from grid distances are made during the process of the computations so that a map of the parcel can be produced that shows the grid bearings of the lines, the level surface distances between the corners, and the level surface area of the parcel.

It is important to note that the preceding example produced quantities for directions and distances that were based upon a series of measurements that had been repeatedly adjusted in an attempt to account for errors. The condition of the corner stones was such that an interpretation was required to determine the former, upright location of the stones. If a second surveyor were to clear the line between two of the stone corner monuments of the example parcel and directly measure the distance between them, the distance reported as a result of the direct measurement would be expected to vary slightly from the distance reported by the first surveyor. Moreover, if a third surveyor were to measure the same distance between the same two monuments, directly, the results of that third measurement would vary slightly from the others. If anything is learned from this example, it is that angles and distances reported on all survey plats and maps are the results of measurements and are never absolutely "correct."

The modern survey, as in the preceding example, has reduced the uncertainties of the angles, distances, and areas reported by the surveyor to a minimum. Compared to the methods used during the late nineteenth and early twentieth centuries, when most of the titles of private property were developed, the modern survey is vastly more accurate and, consequently, more reliable. The laws governing real property transfers recognize the uncertainty of measurements made by surveyors and operate effectively in spite of these ancient deficiencies.

The future of distance measurement is uncertain, but certain advanced concepts have begun to emerge. The use of specialized transmitters in earth orbit to determine locations has been used to a limited degree, with promising results. Known as global positioning satellite (GPS), this technology may prove to be the method of the twenty-first century. GPS is, in some ways, still in the experimental stage, yet many argue that it is ready for practical application and general acceptance by professional surveyors. Implementation of the GPS requires the use of geodetic positions and is most effective when geodetic positions are translated into state plane coordinates.

This short history of measurement systems emphasizes the third,[5] and most compelling, reason for the canon that boundaries are defined by the location of the corners. *If boundaries were defined by directions and distances, every advancement in the science of measurement would move boundaries!*

6.12. ELEVATIONS IN THE MODERN PERIOD

Surveyors have always been aware of the role that changes in elevations (vertical distances) play in measuring and mapping real property parcels. Until very recently, the interest in the elevation of a property parcel was rarely associated with a national datum. Landowners were only interested in relative elevations. Was the property hilly or flat? Was the land subject to frequent flooding? These and other questions relating to relative elevation were rarely part of the real property deed documentation and were usually reserved for the inquiry of sophisticated buyers.

The process used to measure the differences in elevation has changed little in the last century, in spite of all of the other advances in technology. The greatest change has been one of regulation, not science. Federal, state, and local regulatory authorities have intervened in the process of real estate transactions. Federal flood insurance programs, lending agency policies, state and local regulations have added requirements for information that go far beyond the traditional documentation of real property parcels. The absolute

---

[5]The first reason that corners define boundaries is ancient tradition. The second is the need for regular shapes on an irregular planet.

elevation of structures relative to a specific datum is now a common requirement of many regulatory agencies.

Before an agency can require elevation data pertaining to a real property parcel, that agency must establish or specify the datum to be used. The federally defined NGVD (section 3.8) has become almost universal as the datum of choice for all regulatory bodies. Many regulatory agencies refer to NGVD elevations as a requirement without understanding the least bit what is involved. The process of measuring the NGVD elevation of a structure and the foundation of that value need to be understood if that information is going to be used in a beneficial way.

The National Geodetic Survey (NGS) has established a series of "bench marks," which are simply specific locations that have a published elevation associated with them. A typical bench mark is a brass disk clamped onto the end of a very long rod (often 150 feet long). The rod is driven into the ground or otherwise fixed to the earth at what is hoped to be a stable site. The differences in elevation between all of these bench marks are measured using differential leveling procedures that are a sophisticated version of the surveyor's ordinary level and leveling procedure. These differences in elevation are measured and remeasured frequently. New bench marks are constantly being added, and old bench marks are frequently being destroyed or disturbed. Furthermore, the entire network of bench marks is constantly being re-evaluated and updated.

The very definition of the NGVD is scheduled to change very shortly from the "datum of 1929" to the "datum of 1988." This regulatory change will have a small impact in some areas and a significant one in others. A location having an elevation of "x" NGVD '29 will be redefined as having an elevation of "y" NGVD '88. Federal, state, and local regulations will have to be amended to reflect this change in the definition of NGVD.

The most common method used by the private surveyor in determining the NGVD '29 or '88 elevation of a location is differential leveling, using a surveyor's level and level rod. The modern surveyor's level is a device fixed with a telescopic sight that can be mounted on a tripod and adjusted so that the line of sight through the telescope at the center cross hair is perpendicular to the plumb line at the instrument. The level rod is a graduated ruler that, when set vertically on a surface, displays the vertical distances along the rod from that surface.

The process of differential leveling begins at a bench mark or other location of published elevation. The level is set up a short distance from the bench mark (usually less than 150 feet), and the level rod is set vertically on the bench mark. The observer then peers through the level's scope and records the value seen on the level rod at the horizontal line of a center cross hair. This value is the measured distance between the surface and the line of sight. If the earth were flat, the value seen on the rod would also be the difference between the elevation of the level and the elevation of the bench mark. The elevation of the level would equal the elevation of the bench mark plus the rod reading.[6]

The rod is then moved to a new location nearer the site to be surveyed (still less than 150 feet from the level) and set vertically on a new surface. The observer again records the value shown on the rod through the sight of the level. This second value is subtracted from the just determined elevation of the level, producing (theoretically) the elevation of the second surface. The level is then moved ahead, and the process is repeated, using the newly determined elevation of the second surface or "turning point" to compute the elevation of the level at the new location. This process is repeated, in leapfrogging fashion, until the location to be measured is reached. This laborious and tedious process of differential leveling requires at least 17 instrument setups to "carry" elevations 1 mile. The network of level runs needed by NGS to establish the federal bench marks consisted of hundreds of thousands of miles!

Unfortunately, elevation information, so laboriously collected, is not permanent. Aside from the changes in elevations wrought by redefining the datum, many regions are experiencing vertical movement in the land itself. Heaving and subsidence of the earth's surface is a common phenomenon in many areas. The elevation of a building is not a constant value but changes with time. Even in areas where there is very little vertical movement, the significance of an elevation value will change. Land development will change the storm water runoff patterns, often subjecting land to an increased frequency of flooding. The maps used by federal, state, and local agencies to determine flood hazard are often updated and revised, based upon changes in the watershed. Elevation information must be current to be of any value.

---

[6]The world isn't flat, of course. For that, and other reasons, the distance from the level to the rod is kept so small. The myriad complications associated with precise differential leveling is far beyond the scope of this book.

CHAPTER 7

LAND RECORD SYSTEMS

Many territorial animals identify land parcels by marking the boundaries with their scent, scrapes, or other physical signs that tell other animals, "This is mine—keep out." The human race is no different from other territorial animals in this respect. The method of marking boundaries most frequently used by humans consists of marking boundary corners with physical objects, such as posts, rocks, pipes, or iron rods, and marking the lines between the corners with fences, ditches, roadways, or tree lines. This is fine for demonstrating to others the physical limits of a parcel of land. This is not adequate, however, when the need arises to define or describe that parcel to others who cannot visit each boundary corner.

7.1. THREE BASIC TYPES OF LAND RECORD SYSTEMS

The private ownership of land and the statutes that regulate the transfer of that ownership require that each and every separate parcel, public or private, be capable of being uniquely distinguished from all other parcels in writing. The systems of identification used in each state or region vary but usually take some form of one of the two basic systems of identification for original tracts. These two basic systems are the "metes and bounds" and the rectangular United States Public Lands System (USPLS). In many places, or because of interstate ownership of property, variations of both systems are used simultaneously. A third system of parcel identification, the plat or lot and block identification (platted subdivision), is used in cases where original tracts have been subdivided into smaller parcels or lots. Each system will be described in detail.

The first system to be employed in the United States was the "metes and bounds" system and is the basis for property identification in most areas settled and in private hands before the Revolutionary War. This system originated at the dawn of civilization and grew to favor in Europe during the eighteenth century, when the increase of privately owned (and taxed) land prompted governments to develop more accurate cadastre records. The metes and bounds system requires that privately owned land be identified by naming the adjoining owners and physical limits (bounds) of a property, and that a report of the dimensions (metes) be included in the description of the property.

This system of identifying or describing real property has often been erroneously referred to as the "legal description" when, in fact, any proper description based upon any of the systems of identifying land parcels is a "legal" description. The additional requirement that all land transactions be in writing and recorded with some governmental body to assist the local tax collector in developing a record of property owners (and tax debtors) meant that a general description of the property that simply named all adjoiners (bounds) was insufficient.

The second system, which affects the majority of the land area in the United States is the United States Public Land System (USPLS). The USPLS is a more regulated, formal system of identifying land parcels than the old European metes and bounds system. In 1784, a congressional committee, chaired by Thomas Jefferson, proposed a rectangular division that became the Ordinance of 1785 and the foundation of the USPLS. It is very important to remember that the prime purpose of the USPLS was to divide U.S. public land quickly into easily and uniquely identified parcels so that they could be sold and taxed. Although based upon field surveys, the precision (accuracy) of these surveys was of the lowest priority.

The third system, used throughout the United States, comes into play where large tracts have been subdivided into several smaller lots. In these cases, reference to the lot and/or block numbers or letters shown on a "subdivision plat" is the means by which a land parcel is identified. The subdivision plat is a plan, usually recorded, that specifies the location and dimensions of several parcels of land. A subdivision plat may divide a tract of land that was originally defined by metes and bounds or by the USPLS.

In spite of the possibility that any one of the preceding systems for identifying land is sufficient for sale and tax purposes, the tendency throughout the United States has been to add to or supplement USPLS and plat descriptions with a metes and bounds description for each sale of property. This dual use of property descriptions has led—can only lead—to instances of confusion in cases where the two descriptions conflict. Much of the confusion could be eliminated if land title users would realize that metes and bounds descriptions of USPLS or platted parcels were intended to *supplement* the description, not *supplant* it.

7.2. METES AND BOUNDS SYSTEM

The first settlers of the New World began to claim, occupy, and divide up parcels of land based upon the European system of defining or describing land according to its relative position to well-known landmarks and by identifying the adjoining landowners. Although the large land settlement companies usually claimed land based upon latitudinal boundaries, the precision needed to define smaller, individual holdings by astronomic position was not possible. It was much simpler, quicker, and more distinctive to choose natural boundaries to designate the limits of large tracts. Rivers, being the transportation arteries at the time, figured heavily in these early boundary definitions. Other natural features were also used and were usually chosen for their distinctive appearance or durability.

Of course, it was difficult to find rivers, streams, or rock formations in just the right locations to use as landmarks when the limits of a parcel of a particular size were being defined. This meant that artificial landmarks or monuments had to be set by the owners as a substitute for natural ones. Again, the limits of the parcel were defined by the actual location of particular corner landmarks or monuments.

In many cases of areas of long settlement, the construction of fences, hedgerows, or other obvious physical signs of the parcel limits made the task of marking the boundaries less difficult. The primary means of distinguishing one parcel from another became the ''bounds'' system. Land was held and worked by individuals on a regular basis. One need only name the individuals who owned, and usually lived on, the land surrounding a particular parcel to define where it was and what were its limits. This ''bounds'' system did not require that the size of the parcel be determined in order to define it.

In order to describe *roughly* the size of land parcels and assist in defining the location of that parcel, the concept of ''metes'' was used, along with the naming of natural landmarks, artificial monuments, and adjacent owners. The business of measuring between corners, outlined earlier, was taken on by anyone and everyone. The corners or limits of the land were physical marks on the ground. The ''metes'' were the measurements between those marks. Often a landowner would simply pace off a section of land, set boundary stones, and sell the land so marked to another. The stones, neighboring landowners, and the paced distances would all be recorded in the sale, but the land would be occupied, cultivated, and possessed according to the location of the *corner monumentation*. This record of the distances (metes) and *physical* limits (bounds) of a land parcel constituted a "metes and bounds" description.

The metes and bounds description of a land parcel is, in essence, the words that draw a picture. It takes the form of an imaginary trip, leading the reader from a relatively well-known landmark to and around the parcel being defined. The naming of landmarks or monuments at each corner, along with the identification of each adjoining parcel, is an essential part of the description. The metes and bounds description is so ingrained in the land title community and dovetails so well with the laws regulating the transfer of titles that, even in areas where more precise land identification systems are used, the metes and bounds description is often used to supplement descriptions. A complete boundary description using the metes and bounds system will incorporate the following:

1. A commencing point that is well known, easily found, permanent, recoverable, recognizable, and preferably public in origin. The purpose of the commencing point is to clarify the location of the parcel in relationship to some universally recognized landmark.

2. A point-of-beginning that is a part of the property being described. The purpose of the point-of-beginning is to emphasize that the limits of the parcel itself are to follow in the dialogue. The point-of-beginning is used purely for descriptive purposes and is *not* the point where a surveyor "begins" a survey.

3. A report of the physical objects (monuments) that mark the location of ends, and sometimes areas, along each line.

4. A report of the contiguous owners along each line or land record identification of the contiguous parcels (bounds).

5. A direction, usually a bearing, for each line. In some cases, the angle formed at a corner may be reported in lieu of a bearing. In the cases of curved or meandering boundaries, the appropriate words necessary to describe the configuration of the line are used.

6. A distance between each corner (metes).

7. The area of the parcel.[1]

8. Reference to the particular plat or map of survey that forms the basis for the description.[2]

---

[1] The area is not a necessary part of a metes and bounds description.

[2] All metes and bounds descriptions must be based upon a survey. The survey may have been crude, unplatted, and performed by the individuals involved in the sale, but, unless someone

Metes and bounds descriptions are, of necessity, redundant. A particular boundary will have several requirements placed on it. A boundary line will be reported as having a particular adjoiner, having certain monuments as its limits, having a certain length and direction, and as combining with all of the other boundaries to form an enclosed parcel of a particular area. It is not difficult to see that, given the inaccuracies of measurements, many of these elements might be contradictory. The history of the development of the metes and bounds description shows that, for the most part, the physical limits of a parcel controlled possession and use of the land. Because of these factors, the practice, in most areas, has settled upon a "hierarchy of calls" that rank the elements of a metes and bounds description. This ranking of calls is based upon the rules of evidence and is commonly broadened to assist in interpretation of all forms of descriptions and boundary recovery.

The most common hierarchy of calls is as follows:

Natural monuments[3]

Artificial monuments[4]

Directions

Distances[5]

Areas

Any element of a description of a parcel may be rejected or overruled, based upon a review of the best evidence available. The courts have usually held that the most important and overriding factor in the interpretation of property descriptions is *the intent of the parties.*[6] All of the words of a deed are to be considered so that evidence that best demonstrates the intention of the buyer and seller will prevail, in most cases. With this in mind, the writer of a metes and bounds description should ensure that the intention of the parties is clearly expressed.

"measured" the distances and visited the bounds, a metes and bounds description is impossible.

[3] Natural monuments may consist of rivers, rock formations, trees, and other distinct features. Features made by humans, such as canals, levees, highways, or mounds, may be considered as natural monuments under certain conditions.

[4] Objects placed for the specific purpose of marking boundary corners, such as posts, concrete posts, iron rods, pipes, or other physical markers, constitute artificial monuments only when they are correctly set and undisturbed. Many surveyors' markers represent locations that are not boundary controlled. These markers are not "artificial monuments" in the context of this listing.

[5] In Louisiana, the status of distances and directions is reversed. In some states, these items are combined.

[6] C. M. BROWN, W. G. ROBILLARD, D. A. WILSON, EVIDENCE AND PROCEDURES FOR BOUNDARY LOCATION (1981).

The significance of this ranking will be evident during the examination of the process of corner recovery. The charge that society gives to the land surveyor is to recover the boundaries where they exist on the ground. The wonderful precision of measurement and location that is possible today is of little use if an incorrect location for a boundary corner is used. A poor measurement to the correct corner is much superior to a precise measurement to a false corner. In real property boundaries, it's the physical location of the corners that is important.

During the early years of the settlement of the United States, when metes and bounds descriptions were the primary system of identifying and locating land parcels, the fact that every legal document had to be handwritten limited the use of survey plats to sketches and caused many to shorten descriptions to the point that the intentions of the parties became obscure. The dimensions of the parcels were often ignored or reduced to a report of adjoiners only.

Often the dimensions reported in these early descriptions were the results of estimations by the landowners, not surveyors. Measurements by amateurs, and indeed some surveyors, most commonly report distances between monuments as being greater than the actual distance. Rough terrain, poor chaining techniques, meandering routes, and other error-developing factors more often misled all into believing a greater distance was covered than was actually traversed. Because vast areas of land were rapidly being settled, the shortage of skilled surveyors forced many to depend upon themselves to measure their land. Even today, in order to save a few dollars, many land transfers take place without a survey to verify the dimensions of a parcel.

The situation of possessing according to physical marks should be distinguished from the case of a transfer of property where relying on specific dimensions is the intention of the parties. If a sale is not followed by possession and it was the clear intent of the parties to transfer a land parcel of specific size and shape, then any blunders in marking the limits of such a parcel can be corrected at any time. The intent to transfer a specific-sized parcel can only occur in cases of resubdivision. Original private claims and USPLS parcels cannot fall into this category, because *possession* was required in the first case and the *corners set by the government surveyor* controlled in the latter.

7.3. READING A METES AND BOUNDS DESCRIPTION

The most difficult part of writing a metes and bounds description is deciding the degree of detail required to convey the intentions of the parties. The natural instincts are to limit the verbiage as much as possible so that the resulting descriptions are not overly "wordy." Unfortunately, this trend has led to many misinterpretations in the past. Instead of rehashing the general rules, perhaps a review of some typical parcels and how their metes and bounds descriptions might be created would better clarify the need for specificity.

7.3.1. The Johnston Property

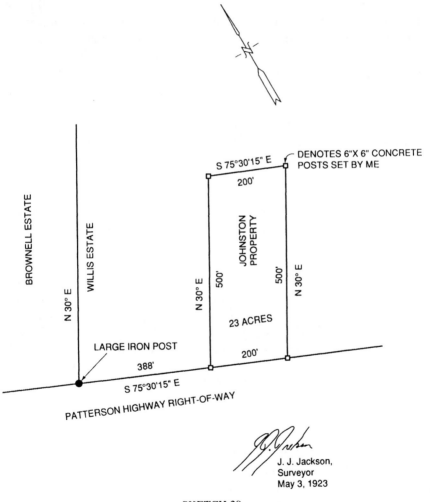

SKETCH 28

In Sketch 28, the iron post shown is reported by the surveyor as monumenting the intersection of the line between the parcels known as the "Brownell" estate and the "Willis" estate with the northernmost right-of-way line of "Patterson Highway." Each of the corners of the "Johnston" property is monumented by 6-inch square concrete posts. The Willis estate, in this example, is the parent tract from which the Johnston property was formed. It is the intention of the parties that the parcel sold be 200 feet wide, 500 feet deep, commence 388 feet from the western boundary of the Willis estate, as measured along the highway, and that the sidelines of the parcel be parallel with the western boundary of the parent estate.

The plat of the survey by Jackson clearly reflects these intentions and provides some additional information. The large iron post is a well-known (in 1923) and accepted feature that qualifies as an artificial monument. The 6-inch square concrete posts are simply survey markers set by Jackson. These may mature into artificial monuments, provided sufficient acts of possession and reliance on these posts take place. There is a discrepancy on the survey plat in the form of an incorrectly computed area. Assuming that all of the other dimensions are correct, the area should be 2.21 acres (rounded off to the nearest one-hundredth of an acre).

A typical metes and bounds description of the period might read:

> . . . a 2.3 acre parcel of land measuring 200 feet wide by 500 feet deep between equal and parallel lines and beginning 388 feet from the western boundary of the Willis Estate, bounded in the front by the Patterson Highway and on the sides as well as in the rear by the property of the seller. . . .

Note the lack of reference to the survey plat by Jackson. If such a reference had been made, then the several discrepancies between the description and the intention of the parties might be clarified. Even if such a reference were made, there would always exist the possibility that the survey plat might not survive. For this reason, metes and bounds descriptions should be written as if all other documents reflecting the intentions of the parties were going to be destroyed.

Sketch 29 shows one interpretation of the description superimposed upon the original survey plat. The scale is distorted to emphasize the discrepancies.

The description does not reflect the intentions of the parties to the sale. Later surveys that recover the concrete posts set by Jackson will require additional evidence and be accompanied by sufficient possessive acts before it can be stated with confidence that they do indeed represent the intended limits of the Johnston property. Even if the survey plat by Jackson were to surface, it might be argued that it was not relevant because it was not part of the written instruments included in the transaction. Now consider the following description:

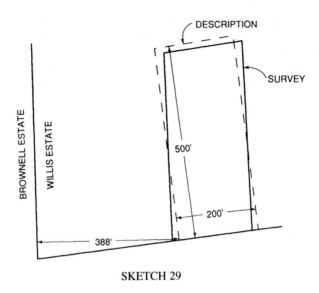

SKETCH 29

. . . commencing at the large iron post[7] found at the intersection of the boundary line between the Brownell Estate and the Willis Estate and the northernmost right-of-way line of Patterson Highway[8]; thence, in an easterly direction[9] along said northernmost right-of-way line of Patterson Highway, [10] South 75 degrees 30 minutes 15 seconds East, a distance of [11] 388 feet to the Point-of-Beginning[12]; thence, in a northerly direction parallel with the boundary line

[7] Reference to the "iron post" fixes the commencement point to a recoverable terrain feature. If, at some later date, the location of the boundary between the Brownell and Willis estates were to become unclear, the location of the Johnston property would not be cast in doubt.

[8] The reference to the intersection of the boundary line and the right-of-way line of the highway is used not only to document the location of the iron post, lest it be disturbed, but also to emphasize that the right-of-way for Patterson Highway is relevant to the measurements.

[9] The words "in an easterly direction" are used to assist the reader in interpreting whether the south–east or the north–west version of the bearing that follows is appropriate.

[10] This removes any doubt about the line of measurement of the 388 feet. It is clear that it was not the intention of the parties that the 388 feet be measured at right angles to the boundary line.

[11] The use of "a distance of" assists the reader in distinguishing distances along the described boundary line and other dimensions that maybe necessary, such as a radius for a curve, and so on.

[12] If one were to insert the phrase "to a 6-inch square concrete post" here, then the distance "388 feet" might be interpreted as an estimation and the concrete post might be considered an artificial monument, without any acts of possession. Blunders by Jackson in setting the posts may not be correctable. Adequate possession with reliance upon the post would make this a moot point.

between the Brownell Estate and the Willis Estate, [13] North 30 degrees East, a distance of 500 feet; thence, in an easterly direction parallel with the northernmost right-of-way line of Patterson Highway, South 75 degrees 30 minutes 15 seconds East, a distance of 200 feet; thence, in a southerly direction parallel with the boundary line between the Brownell Estate and the Willis Estate, South 30 degrees West, [14] a distance of 500 feet to the northernmost right-of-way line of Patterson Highway; thence, in a westerly direction along said northernmost right-of-way line of Patterson Highway, [15] North 75 degrees 30 minutes 15 seconds East, a distance of 200 feet to the Point-of-Beginning, encompassing an area of 2.3 acres [16] and all as more fully described on a plat of survey by J. J. Jackson, Surveyor dated May 3, 1923 [17]. . . .

This description is somewhat longer than the first example, but it presents a much more accurate picture of the transaction. The inclusion of the survey plat as part of the description also gives a clue to the source of the information given in the description.

Sketch 30 represents a survey plat of the same Johnston property 55 years later. The changes that time has wrought are not as great as might have been expected. Two of the original concrete posts are still standing and, from all indications, were not disturbed during all that time. It is known that Abraham Johnston built the house in 1926 and has lived there ever since. The acceptance and use of these concrete posts by the possessor of the parcel has converted them from mere survey markers, attempting to demonstrate the intent of the parties, into artificial monuments that clearly define the limits of the parcel.

The easternmost line of Comstock Road is believed to be the original line between the Brownell and Willis estates. So many changes have occurred that a new metes and bounds description is required in order for the property to be recognized. The application of the hierarchy of calls, as well as the

---

[13] By calling for the sidelines to be parallel to the western boundary, any measurement blunder by Jackson should not alter the location of the line.

[14] Notice that the survey plat indicates the direction to be "North 30 degrees East." The survey plat is a two-dimensional drawing; therefore, "North 30 degrees East" and "South 30 degrees West" define the same line.

[15] This leaves no doubt that frontage on Patterson Highway is intended for this parcel.

[16] The incorrect area is repeated in the description but is of no consequence in the absence of any evidence that it was the controlling factor in the intention of the vendor or vendee.

[17] The reference to the survey plat in the description has the effect of including the plat as an integral part of the description. Discrepancies between the written description and the figures on the plat are normally resolved in favor of the plat, unless clear and convincing evidence to the contrary is discovered.

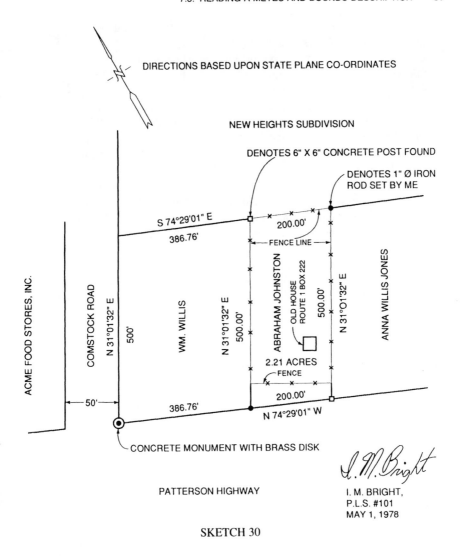

DIRECTIONS BASED UPON STATE PLANE CO-ORDINATES

NEW HEIGHTS SUBDIVISION

DENOTES 6" X 6" CONCRETE POST FOUND

DENOTES 1" Ø IRON ROD SET BY ME

S 74°29'01" E
386.76'

200.00'

FENCE LINE

ACME FOOD STORES, INC.

COMSTOCK ROAD

N 31°01'32" E

500'

WM. WILLIS

N 31°01'32" E
500.00'

ABRAHAM JOHNSTON

OLD HOUSE ROUTE 1 BOX 222

500.00'

N 31°01'32" E

ANNA WILLIS JONES

2.21 ACRES
FENCE

200.00'

N 74°29'01" W

50'

386.76'

CONCRETE MONUMENT WITH BRASS DISK

PATTERSON HIGHWAY

I. M. BRIGHT,
P.L.S. #101
MAY 1, 1978

SKETCH 30

professional judgment of Bright will be noted during the development of the updated metes and bounds description.

. . . commencing at a brass disk set in concrete found at the intersection of the easternmost right-of-way line of Comstock Road and the northernmost right-of-way line of Patterson Highway[18]; thence, in an easterly direction

---

[18] This commencement point may or may not be in the identical location as the large iron post found in the 1923 description. The differences in the commencement distance that follows may be the result of a slight change in the accepted position for the line between the original estates or may indicate a measurement discrepancy between Bright and Jackson.

along said northernmost right-of-way line of Patterson Highway, South 74 degrees 29 minutes 01 seconds East,[19] a distance of 386.76 feet to a 1 inch diameter iron rod on the easternmost boundary line of the property of William Willis and the Point-of-Beginning[20]; thence, in a northerly direction along said easternmost boundary line of the property of William Willis,[21] North 31 degrees 01 minutes 32 seconds East, a distance of 500.00 feet to a 6 inch square concrete post on the southernmost boundary line of New Heights Subdivision[22]; thence, in an easterly direction along said southernmost boundary line of New Heights Subdivision, South 74 degrees 29 minutes 01 seconds East a distance of 200.00 feet to a 1 inch diameter iron rod on the westernmost boundary line of the property of Anna Willis Jones; thence, in a southerly direction along said westernmost boundary of the property of Anna Willis Jones, South 31 degrees 01 minutes 32 seconds West a distance of 500.00 feet to a 6 inch square concrete post on the northernmost right-of-way line of Patterson Highway; thence, in a westerly direction along said northernmost right-of-way line of Patterson Highway, North 74 degrees 29 minutes 01 seconds West, a distance of 200.00 feet to the Point-of-Beginning, encompassing an area of 2.21 acres and all as more fully described on a plat of survey by I. M. Bright, P.L.S. dated May 1, 1978. . . .

In the day of hand-copied records and manual typewriters, the preceding description would have been condemned as much too verbose. In the modern world of copying machines and word processors, there is no need to sacrifice clarity of intent by reducing the length of a metes and bounds description. Some may still insist that the exact same idea can be expressed in fewer words. If we trim any mention of "New Heights Subdivision," then the description would read as follows:

. . . commencing at a brass disk set in concrete found at the intersection of the easternmost right-of-way line of Comstock Road and the northernmost

[19] The route of the description is in an easterly direction. Therefore, the bearing must be south–east, not north–west. The difference in magnitudes is not significant. The modern survey is oriented to state plane directions; the 1923 survey does not report a bearing base.

[20] The call is to the iron rod in this case, because nearby original concrete posts enabled Bright to recover and remonument the southwest and northeast corners of the tract. The original concrete posts have clearly become artificial monuments, marking with certainty the limits of the Johnston property.

[21] There is no longer the need to specify the line as parallel to the original boundary between the Brownell and Willis estates.

[22] The fact that the distances are now reported to the one-hundredth of a foot is an indication of the precision of the measurements. The size of the end monuments (6inches × 6 inches and 1inch o.d.) are such that a variation of 0.10 feet or so ought to be expected. Bright chose to report the same metes as Jackson.

right-of-way line of Patterson Highway; thence, in an easterly direction along said northernmost right-of-way line of Patterson Highway, South 74 degrees 29 minutes 01 seconds East, a distance of 386.76 feet to a 1 inch diameter iron rod on the easternmost boundary line of the property of William Willis and the Point-of-Beginning; thence, in a northerly direction along said easternmost boundary line of the property of William Willis, North 31 degrees 01 minutes 32 seconds East, a distance of 500.00 feet to a 6 inch square concrete post; thence, in an easterly direction, South 74 degrees 29 minutes 01 seconds East a distance of 200.00 feet to a 1 inch diameter iron rod on the westernmost boundary line of the property of Anna Willis Jones; thence, in a southerly direction along said westernmost boundary of the property of Anna Willis Jones, South 31 degrees 01 minutes 32 seconds West a distance of 500.00 feet to a 6 inch square concrete post on the northernmost right-of-way line of Patterson Highway; thence, in a westerly direction along said northernmost right-of-way line of Patterson Highway, North 74 degrees 29 minutes 01 seconds West, a distance of 200.00 feet to the Point-of-Beginning, encompassing an area of 2.21 acres and all as more fully described on a plat of survey by I. M. Bright, P.L.S. dated May 1, 1978. . . .

At first glance, there doesn't appear to be any difference between the two descriptions. Let us assume that the second version is used and that the northernmost concrete post, as well as the iron rod, are destroyed. Later, when New Heights subdivision's southern boundary is monumented and found to be 501 feet from Patterson Highway, who owns the strip between the 500-foot rear line of the Johnston property and New Heights subdivision? The boundary action needed to settle this aggravating, petty discrepancy might have been avoided by the use of a complete, albeit wordy, metes and bounds description.

7.4. U.S. PUBLIC LAND SYSTEM (USPLS)

I know of no place where the quote by Will Rogers, ''The trouble ain't so much what we don't know—its what we do know that just ain't so,'' is more applicable than in USPLS states. A close examination of the origin and history of the USPLS will lead us to a better appreciation of the ingenuity—and pitfalls—associated with this system.

At the end of the Revolutionary War, the United States was a new nation without a firm revenue base. Taxes were unthinkable; yet it takes money to operate a federal government. The one thing that the fledgling country did have was land. From the original state cessions in 1781 to the purchase of Alaska in 1867, approximately 1,807,682,000 acres of land were acquired by the federal government, and the faster this land could be turned into hard cash, the better. The prime, yet unwritten, directive to the committee that created the Ordinance of 1785 was to divide public lands into marketable portions. In order to accomplish this, the following things, listed by priority, were necessary.

1. The parcels had to marked on the ground so that the new landowners could take possession of their land.
2. The parcels created had to be uniquely identifiable.
3. The system of identification had to be one that required little formal education to use, if not understand.
4. Further division of the created parcels had to be simple, universal, systematic, and based upon the location of the monumented corners, and could not require additional survey work by the government surveyor.
5. The time and paperwork required to divide each parcel had to be kept to a minimum.
6. The system could not depend upon prior knowledge of the terrain in order to implement it.
7. The quality or arability of the land had to be reported to determine relative value.
8. The dimensions of the sides of the parcel and other measurements had to be reported to *facilitate the recovery of the corner markers,* as well as estimates of acreage.

Notice the priority given to measuring the dimensions of the newly created parcels. Land lines are defined by the corners *set by the government surveyor,* not by the reported dimensions or directions. The USPLS fulfilled these requirements completely and in the required order of priority. It is important to keep this order of priority in mind when dealing with USPLS parcels.

As the USPLS was perfected over the years, slight changes in the techniques, numbering order, monumentation, and other minor alterations took place. The 20 states that do not operate under the USPLS are Connecticut, Delaware, Georgia, Hawaii, Kentucky, Maine, Maryland, Massachusetts, New Hampshire, New Jersey, New York, North Carolina, Pennsylvania, Rhode Island, South Carolina, Tennessee, Texas, Vermont, Virginia, and

West Virginia. The other 30 states are USPLS states, and the exact version of division used in each area vary according to the instructions in force, the surveyor, the terrain, local customs, and other factors.

7.5. INITIAL POINT

As so often occurs in surveying, the theory is simple, but the application is complex. The theory of the USPLS was based upon a rectangular division of land along the cardinal directions. The first step in the division of an area of land, such as a state, under the USPLS, was the establishment of the "initial point" for that division. Although some attempt was made to locate initial points at a specific latitude (a list of initial points was made, reporting both latitude and longitude of the various initial points[23]), the initial point was the actual location, on the ground, as monumented by the government surveyor. The reported position existed only to assist in the recovery of the true location.

7.6. PRINCIPAL MERIDIAN

Having chosen and monumented the initial point, the surveyor established the "principal meridian" by traversing north and south from the initial point. Depending on the instructions at the time, a post or other monument was set at least every 6 miles or as often as every half-mile. The principal meridians were run with particular care, and every attempt was made to run the line "true" north, resulting in an approximation of astronomic north being marked on the ground.

7.7 BASELINE

Lines were also laid out east and west from the initial point and monumented with the same frequency as the principal meridian. This "curved" line is called the "baseline" and is actually a series of short chords between the monuments set by the government surveyor. As with the principal meridian, every attempt was made to follow the "true" or cardinal direction, resulting in an approximation of the astronomic latitude of the initial point.

Sketch 31 demonstrates what should have been the result of the establishment of a typical initial point, principal meridian, and baseline. Additional or "correction" east–west lines may have been monumented at regular in-

---

[23] J. G. McEntyre, Land Survey Systems (1978).

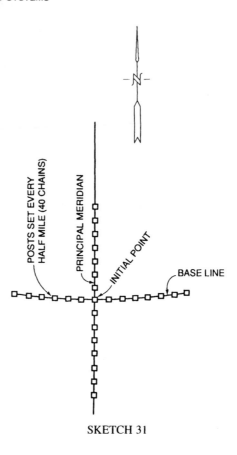

SKETCH 31

tervals along the principal meridian with the same care as the baseline so that a correction could be made for the convergence of the meridians. It was well known by the designers that a rigid rectangular division of land based upon the cardinal directions was impossible, but the inaccuracies of measurement, the scientific unsophistication of the ultimate user of the system, the need for haste, and the need for a simple method of record keeping outweighed the requirements of accuracy.

7.8. TOWNSHIP AND RANGE LINES

After the establishment of the principal meridian and the base line, additional east–west lines were monumented at 6-mile intervals along the principal meridian. These are known as "township" lines. Additional north–south lines were monumented along the base line, and later along correction

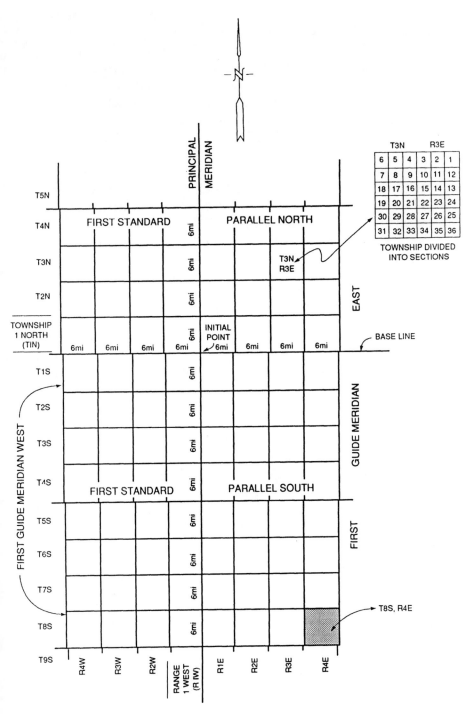

SKETCH 32

lines, at 6-mile intervals.[24] These are known as "range lines." The *theoretical* result of these steps is shown in Sketch 32.

Each east–west row of "6-mile squares" is identified by numbering its position north or south of the base line. Each north–south column of "squares" is identified by its position east or west of the principal meridian. Thus, every square, called a "township," is uniquely identifiable by a simple. report of its township number and range number. Notice that township 8 south, range 4 east has been highlighted in Sketch 32. This is abbreviated to "T8S, R4E"

The township and range lines were not run with quite the same degree of care as were the principal meridian and the base line. Of course, *none* of the townships were exact squares measuring 6 miles on a side. This fact is not important to the division and inventory of the land, and, although not forgotten, it was set aside by the planners of the USPLS because it was considered relatively unimportant.

7.9. IDEAL TOWNSHIP

Further division of each township was accomplished by running lines that were, more or less, north–south and east–west at 1-mile intervals and monumented every half mile. Each resulting "square mile" is called a "section" and identified by a number. The work of monumentation was typically begun at the southeast corner of each township and progressed to the north and west until the entire township was monumented. The discrepancies that resulted from the closure of the meridians, errors of measurement, and undiscovered blunders were accounted for by allowing the western tier and northern row of sections to deviate from the "square mile" template. All other sections were *reported* as squares measuring 1 mile to a side. Of all the over 2 million sections monumented by the Government Land Office (GLO), it is safe to say that *none* is a square measuring 1 mile to a side and few are even close, especially when the precision of modern surveys are considered. Still, the "ideal section" must be discussed in order to explain the USPLS. Sketch 33 is that of the "ideal township," divided into "ideal sections."

7.10. IDEAL SECTION

The division of the "ideal section" into its aliquot parts is something that has been widely sketched and described *as if the dimensions of the sides*

---

[24]The interval between correction lines varied from every fourth to every sixth township line, depending upon the instructions in force at the time.

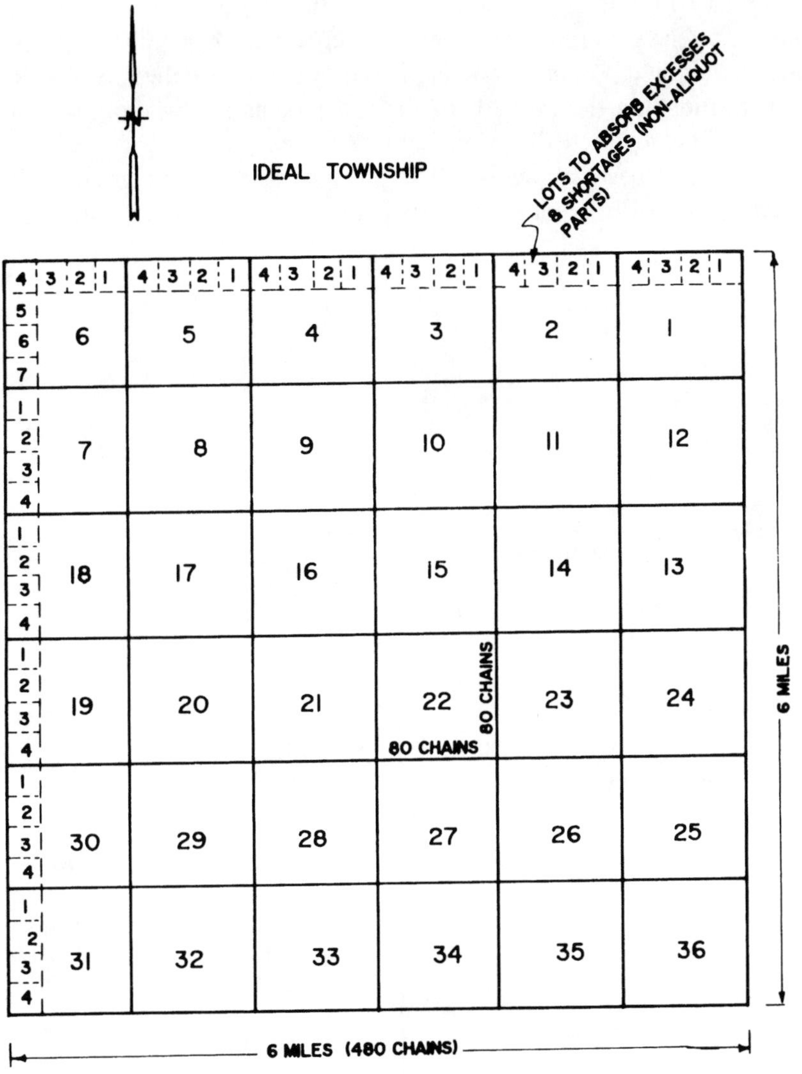

IDEAL TOWNSHIP

LOTS TO ABSORB EXCESSES & SHORTAGES (NON-ALIQUOT PARTS)

4 3 2 1	4 3 2 1	4 3 2 1	4 3 2 1	4 3 2 1	4 3 2 1
6	5	4	3	2	1
7	8	9	10	11	12
18	17	16	15	14	13
19	20	21	22	23	24
30	29	28	27	26	25
31	32	33	34	35	36

80 CHAINS

80 CHAINS

6 MILES

6 MILES (480 CHAINS)

SKETCH 33

were an important part of the process. This is not the case. The beauty of the USPLS is the very fact that it is *independent* of dimensions. It is a system that depends entirely upon the actual locations of the corners set by the government surveyor, and *nothing else*. Sketch 34 is that of the division of a section and how each parcel is identified. The dimensions are not shown, because they are *not essential* to the process and only tend to mislead the novice into believing that sections and their aliquot parts should measure the same as the example in the explanation.

The monuments set by the government surveyor at what was typically reported as 1-mile intervals (80 chains) are the section corners. The monuments set by the government surveyor at, reportedly, half-mile intervals (40 chains) are called "quarter corners." The linchpin of the USPLS is the actual location, on the ground, of the monuments set by the government surveyor. The reported directions and distances can be, and usually are, incorrect or, at best, imprecise. The physical spot on the ground monumented by the government surveyor is, and can only be, one place, nowhere

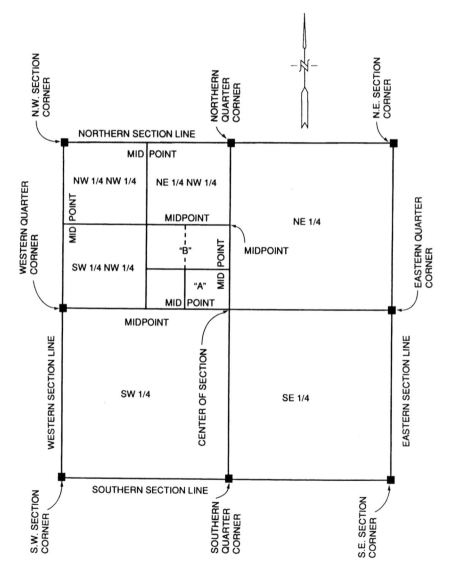

ALL POINTS LABELED "CORNERS" WERE SET BY THE GOVERNMENT SURVEYOR.

SKETCH 34

else. That location can be lost, obscured, misreported, and confused but *never moved*. No matter how imprecise the measurements or how gross the blunders are that may have occurred, the corner monumented can only mark one land point on the earth, and that land point is exactly where it is. Because the actual location of the monuments marking the corners is so vital, extraordinary emphasis must be placed upon the recovery of these locations.

Straight land lines between opposite quarter corners divide the section into a southeast quarter (SE$\frac{1}{4}$), a northeast quarter (NE$\frac{1}{4}$), a southwest quarter (SW$\frac{1}{4}$), and a northwest quarter (NW$\frac{1}{4}$). The intersection of these lines is a land point and is called the "center of the section." Further division is accomplished by connecting straight lines between the midpoints of opposing sides of the larger tract. The description of the resulting parcel is identified by listing the divisions made in the reverse order in which they were made. Tract "A" in Sketch 34 is the southeast quarter of the southeast quarter of the northwest quarter of the section. Tract "B" is the north half of the southeast quarter of the northwest quarter of the section.

If Sketch 34 was that of section 26, in the fourth township south of the base line and the fifth range west of the sixth principal meridian, then tract "A" could be uniquely identified by the single line "SE$\frac{1}{4}$ SE$\frac{1}{4}$ NW$\frac{1}{4}$ Sec. 26, T.4S., R.5W., 6th P.M." Tract "B" would be abbreviated as "N$\frac{1}{2}$ SE$\frac{1}{4}$ NW$\frac{1}{4}$ Sec. 26, T.4S., R.5W., 6th P.M." Note the lack of commas in the list up to the section number. If a comma were inserted in the description of "B," for example, the result might be "N$\frac{1}{2}$, SE$\frac{1}{4}$ NW$\frac{1}{4}$ Sec. 26, T.4S., R.5W., 6th P.M." This could be interpreted as calling for the north half *and* the southeast quarter of the northwest quarter of the section!

The dimensions of tract "A" or "B" in the example are not known and are not important in the description. What is known about the dimensions of these parcels is that, if section 26 were an "ideal" section, with each side measuring 80 chains, then tract "A" would be a square measuring 10 chains on a side and encompassing 10 acres. Tract "B" would be a rectangle measuring 10 chains on the east and west and 20 chains on the north and south and encompassing 20 acres. Of course, section 26 is not an "ideal" section; therefore, tract "A" will not be exactly 10 acres, nor will tract "B" be exactly 20 acres. A recovery and careful survey of the corners set by the government surveyor is required before a more precise knowledge of the dimensions of these parcels is possible.

7.11. IRREGULAR SECTIONS

The pattern of sections, and even townships, had to be interrupted because of several factors, such as existing private claims, large rivers, lakes, state boundaries, national boundaries, the sea, and other irregularities both natural and human made. This resulted in fractional sections, where the reported

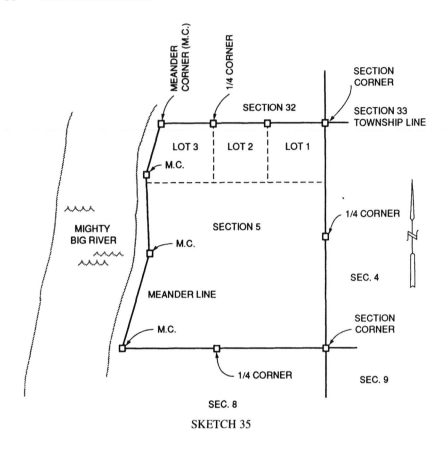

SKETCH 35

configuration deviated from the typical pattern. Aliquot division is still possible in these "irregular" sections by the application of certain rules design to isolate the irregularity.

7.11.1. Meander Lines

Rivers, lakes, the shore of the sea, and other natural barriers that arose in the course of the monumentation of the section lines were accounted for by the "meander line." Sketch 35 illustrates the application of the meander line along a navigable river. The limit of private ownership along a navigable river is defined by the waterway. It is not practical to measure each minute and sinuous turn of the waterway; therefore, a method that approximates the waterway is used.

In the example, the surveyor encountered the river while running west along the line between sections 5 and 8. A meander corner was established on this line *at a convenient point* on or near the bank of the river. Additional meander points were established at points along the river, and the directions and distances between these points were measured so that the approximate shape of the bank of the river could be plotted. This resulted in the line labeled as the "meander line." All distances and directions are to the meander points, but the private ownership extends beyond the meander line to the waterway. Each state has differing laws that define the limit of private ownership along navigable rivers, navigable lakes, the sea, non-navigable lakes, rivers, and streams. These boundaries are known collectively as "riparian boundaries."[25]

Although a general knowledge of the rules of the establishment and division of USPLS lands is important, each section was separately surveyed, monumented, and sold. Because of this, each section is unique in some way. The recovery of the corner monuments and knowledge of the boundaries of any section or its aliquot parts requires special knowledge of that particular section.

7.12. USPLS GENERAL INSTRUCTIONS

Depending upon the year that the work was done, each government deputy surveyor was issued a set of general instructions. These instructions directed the details of measurement, monumentation, observations to be recorded, and method of division, among other things. Subsequent instructions might alter or expand upon previous ones. The instruction told the deputy surveyor how he *should* conduct the surveys; how the deputy surveyor actually performed the surveys is another matter.

7.13. GOVERNMENT SURVEYOR'S ORIGINAL NOTES

Each surveyor kept a record of the distances, directions, observations, monumentation, and other pertinent data in what is called the "surveyor's notes." Typically, these notes were not written as work was progressing but were transcribed from rough notes, usually kept on a slate, at the end of each line. The notes were the best record of what was done, how it was done, the order in which it was done, what was measured, what was encountered along the way, and how monumentation was accomplished.

These notes were used to create the township plats that mapped each

---

[25] Additional rules and definitions dealing with irregular sections and the aliquot division of USPLS sections constitute a book in itself. The reader desiring greater detail is referred to John G. McEntyre, *Land Survey Systems* (New York: John Wiley, 1978).

township, showing the sections and/or lots that were to be sold to private owners. The government surveyor's notes and the township plat could only reflect the directions and dimensions reported at the time of the original survey. The actual configuration of any township, section, or lot was dictated by the location of the corners, not the reported dimensions.

This significant difference is best explained by an example. In Sketch 36, the solid lines represent the reported route of the government surveyor and

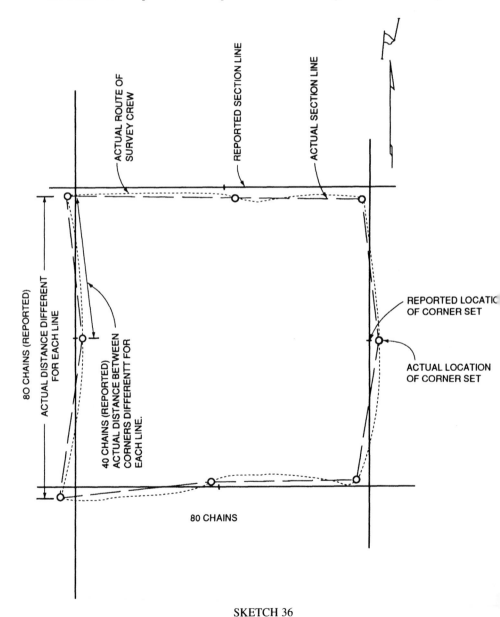

SKETCH 36

the reported location of the monuments set by the surveyor. The small circles represent the actual location of the monuments. The dashed lines represent the actual section lines.

The dramatic differences between the reported and actual locations of the section corners in Sketch 36 are not exaggerations. Variations of that relative magnitude and greater are common, especially in those states of early surveys. The division of the section, being dependent upon section corner and quarter corner location, is accomplished in spite of the inaccuracies of the original reports of direction and distance. As a rule, *all* four quarter corners and at least one section corner *must* be recovered before any aliquot part of a section can be made.

For example, Sketch 37 highlights the SE$\frac{1}{4}$ SE$\frac{1}{4}$ SE$\frac{1}{4}$ SE$\frac{1}{4}$ of the section.

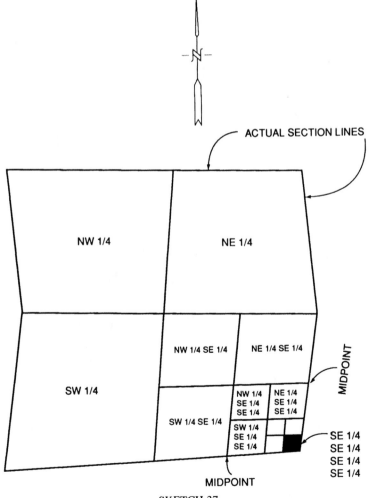

SKETCH 37

If the original survey information were relied upon, one might believe that this parcel formed a square 330 feet (5 chains) on a side and encompassed 2.5 acres. An improperly prepared metes and bounds supplemental description of this parcel might read, in part, ". . . commencing and Point-of-Beginning at the southeast Section corner; thence, West 330 feet; thence, North 330 feet; thence, East 330 feet; thence, South 330 feet to the Point-of-Beginning, encompassing an area of 2.5 acres. . . ."

The novice might argue that one need only recover the southeast section corner, and, by measuring west, north, east, and south the 330 feet called for, this parcel could be monumented. This is simple but quite incorrect, as the very large discrepancies in Sketch 38 clearly demonstrate.

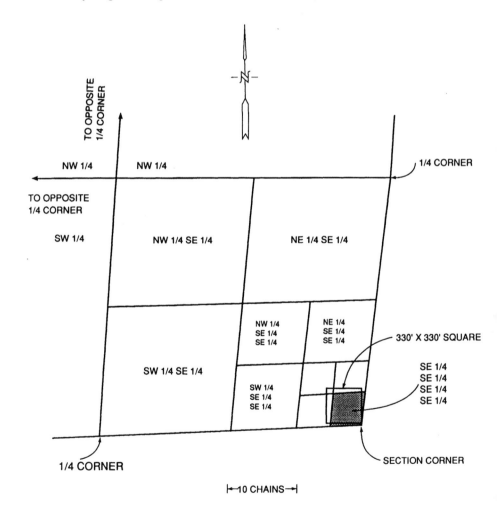

SKETCH 38

In order for the described $SE\frac{1}{4}$ $SE\frac{1}{4}$ $SE\frac{1}{4}$ $SE\frac{1}{4}$ of the section to be monumented, all four quarter corners must be recovered, as well as the southeast section corner. A "straight line" must be run between opposite quarter corners to establish the center of the section. The midpoint of each of the four sides of the resulting southeast quarter must then be set and a "straight line" run between opposite midpoints to establish the center of the southeast quarter. The midpoints of all four sides of the resulting southeast quarter of the southeast quarter must then be established and a "straight line" run between these opposite midpoints to establish the center of the southeast quarter of the southeast quarter of the section. The midpoints of all four sides of the resulting southeast quarter of the southeast quarter of the southeast quarter must be established and a "straight line" run between opposite midpoints to establish the center of the southeast quarter of the southeast quarter of the southeast quarter of the section. Only this procedure will result in the correct monumentation of the described parcel.

Most of the steps that established the midpoints of land lines and the "runs" between opposite sides can be, and almost always are, the results of computations, as opposed to actual field monumentation. The only fieldwork that is absolutely necessary is the recovery of all four quarter corners of the section, the southeast section corner, and the monumentation of the three remaining corners of the parcel. Survey plats or maps of USPLS sections or their aliquot parts that report exactly the same distances and directions as the original government plat are almost certainly the result of incorrect work.

If the reader believes that Sketch 36 offers an overly pessimistic view of a typical section, instead of a regular occurrence, now is a good time to recall the conditions under which most of this work was done. The bulk of the work was done under contracts that paid by the mile. Nineteenth-century work was typically contracted at less than 10 dollars per mile. Out of this fabulous sum, the surveyor had to provide for the payroll and equip the crews with chains, compasses, tents, food, mules, wagons, brush knives, axes, posts, firearms, and other necessities. Whatever remained was the surveyor contractor's profit. It may be possible that some of the surveys were hurried or that some of the work didn't follow the general instructions to the letter. Perhaps there may even be one or more instances of fraudulent field notes being returned to the surveyor general's office.

Considering the wild condition of the land, the rough terrain encountered, the crudeness of the instrumentation and techniques, the hostility of the indigenous population, and the demand for haste, one can only be amazed that any dimensions and directions reported were even close. The wisdom of relying on the actual location of the original monumentation over the purported dimensions becomes more evident as the examination of the physical limitations of the original surveys is reviewed.

7.14. NONALIQUOT DIVISION OF A USPLS SECTION

Care must be taken when regular sections are being divided that the intention to divide along lines that are not aliquot parts is not confused with those that are. The best assurance of this is the creation of a subdivision plat, the assignment of lot numbers to the parcels created, and the recording of the plat. The written instruments that document the sale could stress the intent to deviate from the normal process of dividing regular sections. Finally, all reference to fractions, such as $\frac{1}{2}$ or $\frac{1}{4}$, should be avoided. The use of examples may better clarify situations that dictate deviation from the aliquot methods.

Sketch 39 is of the SE$\frac{1}{4}$ of Section 29, T7N; R14W Somewhere principal meridian. Naturally, there has never been a survey conducted on this quarter section, since the government surveyor set the corners in 1890. Everyone "knows"[26] that this quarter section is 40 chains to a side and contains 160 acres. The dimensions shown in the sketch are the actual dimensions that would be revealed by a modern survey.

The owners have two sons, and they want to give each an equal portion of the property. Suppose number one son is deeded the N$\frac{1}{2}$ SE$\frac{1}{4}$ Sec. 29 and

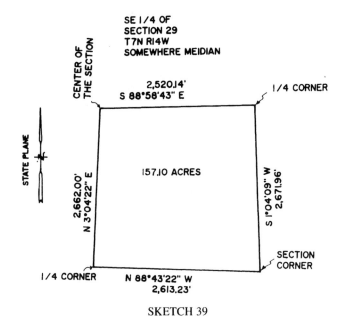

SKETCH 39

---

[26] See page 102 of this text.

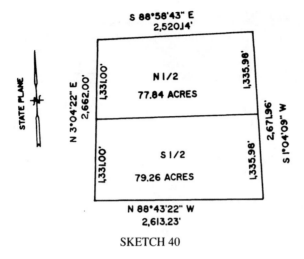

S 88°58'43" E
2,520.14'

N 1/2
77.84 ACRES

1,331.00'

1,331.00'

1,335.98'

1,335.98'

2,671.96'

S 1°04'09" W

N 3°04'22" E
2,662.00'

STATE PLANE

S 1/2
79.26 ACRES

N 88°43'22" W
2,613.23'

SKETCH 40

number two son is deeded the S½ SE¼ Sec. 29. Sketch 40 shows the resulting division.

The division is not equal. The situation could be even more complicated if number one son were to measure south 1,320 feet from the north line of the SE¼, fence the property, and begin cultivation, confidant that he is in possession of his 80 acres. In reality, he possesses 77 acres, because the fence is 10–15 feet north of the correct location. This seems trivial until number two son sells the land to a manufacturing company that builds a warehouse 5 feet from the fence.

7.15. PRIVATE CLAIMS

More often than not, especially in the Mississippi Valley, the government surveyor encountered lands that had been claimed and settled prior to any surveys. In most cases, the private claim was just bypassed, as if it were a large lake. This disruption of the "regular" pattern caused the formation of irregular sections. The section numbering system was adhered to, as nearly as possible, in these cases, with some notable exceptions. The treaty that concluded the Louisiana Purchase required the United States to honor all existing land grants and claims and to conduct surveys of these parcels so that the U.S. Congress could confirm these claims.

This resulted in some very interesting township plats. Surveys of heavily populated areas often produced entire townships where not one regular section occurred. Rarely did the U.S. Congress or the government surveyor make any attempt to rectify pre-existing boundary disputes arising from the

T. 14 S. - R. 25 E. South Eastern District, La.

West of Mississippi River.

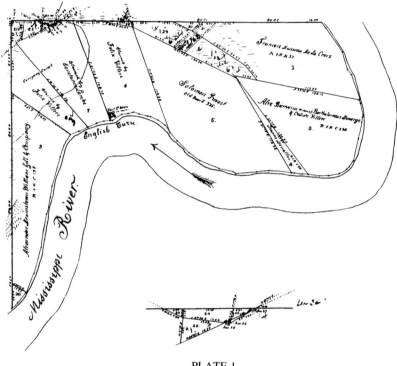

PLATE 1

Table of Contents

Public Land

Sec	Area	Sec	Area	Sec	Area	Sec	Area	Sec	Area	Sec	Area	Sec	Area	Sec	Area	Sec	Area
21	30.9	22	27.64	25	4.56	27	0.01	28	6.60	29	95.68	32	26.66	33	131.08	34	0.32

Total Area of Public Land 323.41 Acres

Private Claims

Sec	Claimants Name	Report Name	Report Date	Report No.	Area	Sec	Claimants Name	Report Name	Report Date	Report No.	Area
1	Francois Duncan de la Croix	R v R	Sept 5th 1833	B. 77	835.79	23	Edmond Fazende and Francois Fazende	R v R	Sept 5th 1833	C. 4	0.66
3	Alix Bienvenu widow of Bartholomew Duverge & Calixte Villere	R v R	"	C. 230	803.62	24	Thaddeus Mayhew	Regr	Nov 6th 1831	19	0.19
4	Jeanne Versailles wife of Pierre Dulcide Baran	"	"	C. 77	77.07	26	Antonio Marcellin Ducros and Casimir Lacoste	R v R	Sept 5th 1833	C 240	0.90
5	Solomon Prevost	Not found		375	1216.00	30	Michel Andry and Hortaire Andry	"	"	C. 56	15.73
6	Claimed by Felix Villere	No confirmation found			507.67	31	Marie Constance Laroche	"	"	C. 30	21.31
7	Claimed by Constance Laroche	"	"	"	500.10	35	Thaddeus Mayhew / Cl. by Constance Laroche	Regr / No confirmation found	1831	19	0.02
8	Claimed by Felix Villere	"	"	"	308.00						
9	Alexander Donaldson, W. P. Hill & Company	R v R	Sept 5th 1833	C 132	770.50	36	Antonio Marcellin Ducros and Casimir Lacoste	R v R	Sept 5th 1833	C. 240	0.14
20	Felix Martin Fontall & Albert Gordon	"	"	C. 2	20.05						

Total Area of Private Claims 5165.58 Acres
" " of Township 5488.99 " "

Traverse of Mississippi River.

Sec	Course	Dist	Sec	Course	Dist	Sec	Course	Dist	Sec	Course	Dist	Sec	Course	Dist	Sec	Course	Dist
1	S.32½ E	7.20	3	S 12½ W	22.68	4	N 86½ W	4.80	5	N 05½ W	12.09	6	S 63½° W	11.36	9	S.19½ W	12.30
"	S.37° E	30.70	"	S.33° W	0.08	5	N.86½ W	25.74	6	N 39° W	8.65	-	N 75° W	1.13	"	S 15½ W	37.45
"	S.17° E	19.14	"	S 43° W	3.00	"	S 86½ W	13.98	"	N 40° W	8.30	7.	S.86 W	6.83	"	S 8° W	11.70
"	S.32¾ E	6.00	"	S 61° W	9.08	"	S.87½ W	22.00	"	N 65 W	10.26	"	S.70 W	8.50	"	S 15¼ W	13.06
"	S.19½ E	20.12	"	S 70¼ W	12.50	"	N.64½ W	2.96	"	N.88° W	2.00	"	S.58 W	2.20	"	S. 10° W	34.11
3.	S.17½ E	2.90	"	S.88° W	9.60	"	N.57¾ W	16.00	"	N.75° W	9.90	-	S.63½ W	8.75	"	S. 15° W	27.04
"	S 6¾ E	15.06	"	N.82° W	13.20	"	N.67° W	23.00	"	N 81½ W	6.80	8	S 64° W	10.12	-	S 37½ W	31.81
"	S.8° W	5.65	4	N.78° W	7.50	"	N.56° W	14.03	"	S 66½ W	5.57	"	S.72¼ W	9.60	20	S.38¾ W	14.02
"	S 10½ W	17.40	"	N.85½° W	8.00	"	N.03° W	10.05	-	S. 86¾ W	7.24	9	S.31 W	13.63	"	S.48° W	8.75

Variation of the Needle 7°57' East

John Maxwell D. S. Surveyed the North Boundary in the 4th qr of 1829 and was paid in the 2d Qr of 1830 he surveyed Sect 1.23.24.26.30 & 31 and all the lines of Public land in the 1st qr of 1830 and was paid in the 3d qr of 1830

William H. Cobb D. S. Surveyed the remainder of Township in the 1st qr of 1830 he was paid for the West Bdy and the Traverse of Mississippi River, in the 2d qr of 1830 but no evidence of payment for the remainder

William Sevey D. S. under contract of 3d March 1853 and Instructions of 7th February 1851 and 12th March 1853 resurveyed the whole Township in the 2d qr of 1853 - The North Bdy. is included in his account approved August 19th 1853

All the sections of Public Land contain Swamp Land accruing to the State of Louisiana under the provisions of an Act of Congress approved 28th September 1850 See List herewith transmitted

Mr Sevey preserved the old boundaries and the old corners wherever found.

Surveyor Generals Office
Donaldsonville La August 19th 1853

The above Township Map has been examined and compared with the approved field notes thereof now on file in this Office found correct and is hereby approved except such stations for which no confirmations have been found

Wm. J. McCulloh
Surveyor General Louisiana.

overlap of private claims and royal land grants. Each confirmed claim was given its own section number, causing some townships to have well over 100 sections.

Plate 1 is a photoreduction of a township plat that was approved by the surveyor general of Louisiana on August 19, 1853. The fieldwork and notes that were used to develop this plat were begun in 1829 and completed in 1853. Note the meander line along the Mississippi River as well as the absence of a regular section. The table of contents reports 5,165.58 acres of private claims and only 323.41 acres of public lands in this township.

Divisions of irregular or "fractional" sections are provided for in the regulations controlling the division of public lands, but such highly irregular sections as are shown in Plate 1 defy any semblance of regular division. In many areas, where such settled conditions existed long before the USPLS implementation, the owners, attorneys, and even the courts continued to use the metes and bounds system of property identification. It is very common, in southeast Louisiana for example, for a real property chain of title to be devoid of any reference whatsoever to section number, township, or range. These areas, although officially in USPLS states, are acting as metes and bounds areas.

7.16. PLATTED SUBDIVISION OR URBAN SYSTEMS

The metes and bounds system and the USPLS discussed so far are systems that are most effective when they are used for large tracts of land. The original land grants of the colonies and the USPLS sections typically transferred land for some agricultural use. Both systems allow for division into very small parcels, but they lose effectiveness or become unwieldy as the size of the parcel decreases. This is especially true for the USPLS. The formal USPLS rules only specify the aliquot division process to the one-sixteenth of a section. The division of USPLS sections into smaller fractions is only an extension of these rules and can become quite clumsy (see Sketch 34).

Very early in the metes and bounds system, and later in the USPLS, landowners often wished to divide their large tracts into smaller parcels. One of the ways in which this was frequently done was to produce a subdivision plan or plat in which the smaller parcels were identified by a lot number or some other identification. This plat was then recorded in the local conveyance office, and any title transfers would refer to the subdivision name and lot number.

This system has many advantages. In urban situations, where public streets are created, the addition of the concept blocks greatly reduces the verbiage needed to describe any particular lot. The plat can be used by the owner as a sales aid. Buyers can readily see on the plat, if not on the ground, the configuration of the property being purchased. The tax collector can use the plat to develop a record of the ownership and relative worth of the properties. Deeds that refer to lot and block identifiers may be supplemented by a metes and bounds description, which refers to street right-of-way corners and adjoining lots to confirm the lot and square.

The platted system of real property identification has one very important distinguishing feature that greatly separates it from the metes and bounds system and the USPLS. Unlike the previous two systems discussed, the platted subdivision does *not necessarily depend upon monumentation or possession* to create boundaries to the property. Before local governments began to exercise their authority, it was not uncommon for owners to draw a platted subdivision of their property solely based upon the old metes and bounds system or USPLS description. It may be that *none* of the corners in the subdivision was ever monumented. It is just as possible that the platted subdivision was properly performed and that all of the newly created corners were monumented. The consequences of both cases will be discussed at length in Chapter 8.

Platted subdivisions should be separated into two additional categories. Subdivisions in which all of the lots were platted and recorded at one time are very different from subdivisions in which the lots were created one at a time. If the former is true, none of the lots can claim "seniority of title."[27] If the lots are created and sold one at a time and some chronological sequence can be established for the transfer of title, then a seniority of title can be established.

Survey markers set at the corners of a platted subdivision do not have the same ranking as corner monumentation by the government surveyor in the USPLS, nor does a survey marker set hold the same status as a corner called for in the metes and bounds description. A platted subdivision usually expresses the intent to convey a parcel of property of specific dimensions. The survey markers, if any are set, reflect a corporeal attempt to place on the ground the limits intended by the survey plat. Any blunder in the setting of these survey marks can be corrected at any time, provided that correction is geometrically possible and sufficient acquisitive acts have not taken place.

Frequently, only the street rights-of-way were monumented by the original subdivider. This created the blocks or squares of the subdivision but did

---

[27] "Seniority of title" refers to the concept that, in the sequential division of land, the parcel sold first must conform, as closely as possible, to the monumentation and dimensions called for in the deed. Parcels sold at a later date must yield to the senior deed in cases of conflict.

not monument the individual lots. Public reliance upon the monumented location of the street rights-of-way will, in most cases, result in any excesses or deficiencies being isolated within each block. It is because of this public reliance that block corner monumentation usually ranks above lot corners in importance during boundary recovery.

In order for a surveyor's marker to mature into the kind of artificial monument that is referred to in the hierarchy of calls, reliance on that marker and the rules of acquisitive prescription must be satisfied. A private surveyor's marker must satisfy several conditions. "Not every cross upon a stone does a corner make, nor each iron pipe a monument" is an amusing but wise phrase.

Unlike USPLS or original private claims, platted subdivisions typically create land parcels that are quite small. Because of their small size, along with the fact that most platted subdivisions are relatively modern, it is more likely that resurveys of individual lots will discover the dimensions to be identical, or at least very close to, the dimensions reported on the original subdivision plat. More often than not, this is because the typically short distances encountered in modern platted subdivisions reduce measurement errors to values that are less than the inherent errors produced by centering on corner monuments.

For example, let us assume that a particular lot in a platted subdivision is reported to be 200 feet from an intersection, 100 feet wide and 200 feet deep. The block corners were originally monumented by 1″ diameter iron pipes and the lot corners by ¾″ diameter iron pipes. A surveyor, recovering the monuments and measuring 200.02 feet and 99.99 feet, respectively, will, in all likelihood, report the original 200 and 100 feet, because the variation from the record is within normal standards of precision.

CHAPTER 8

BOUNDARY RECOVERY

Contrary to popular belief, land surveyors do not establish boundary lines. Boundary lines can only be established by the landowners, although disputes may be settled by adjudication. The act of subdividing a parcel of land by a private surveyor acting as an agent of the property owner, or the monumentation of section corners by the government surveyor in the case of federal lands, is the first step in the act of establishing the boundaries of the newly created parcels that all descendants in title are bound to follow. The land surveyor is charged with the task of recovering those boundaries, not establishing them. A boundary recovery by a land surveyor constitutes a *professional opinion* and is no more binding on the property owners than the advice of any other professional. Every property owner has the right to reject a purported boundary recovery if that owner so desires.

8.1. RULES OF EVIDENCE

Boundary recoveries are opinions rendered based upon the results of an investigation. The rules of evidence that are applied to that investigation are not very different from the rules of evidence in a civil proceeding. Every item discovered during an investigation is evidence. There are no exclusionary rules for a surveyor. Parol evidence, hearsay, private documents, recorded documents, and anything else may be considered or may point the way to further investigation and discoveries. Physical evidence is generally more reliable than testimony, written evidence is given more credence than parol, recorded evidence is given more weight than private, and so on.

The hierarchy of calls (see page 81) used to assist in the interpretation of deed descriptions also guides the surveyor in evaluating boundary evidence. Natural and artificial monuments are physical evidence. A survey plat may be a physical thing, but the measured angles, distances, and computed areas shown on that plat are the recorded observations (testimony) of the previous surveyor. Neighbors and landowners often offer undocumented (and sometimes biased) opinions with regard to the location of boundaries. The surveyor weighs all of the evidence and, based upon the quality and quantity of the evidence, renders an opinion on the location of a boundary.

8.2. SOURCES OF EVIDENCE

Title examinations are usually restricted to recorded documents. The deed document and recorded survey plat are far from the only source of evidence used by the land surveyor in recovering real property boundaries. Because the location of real property boundaries is defined by the location of the real property *corners,* boundary recovery is, in fact, corner recovery. The primary source of evidence is the physical features discovered by the surveyor in the field. Prior to conducting a field search for evidence, the surveyor will use several other sources of evidence in an attempt to narrow the area of search for corner monumentation.

The deed document and previous surveys are obvious sources of evidence. Surveyors who have practiced in an area for a long time will have extensive files that may include deeds and plats of adjacent properties. The county records are also a source of information on contiguous tracts of land. Land surveyors can often be seen looking through the public property records book. They are not looking for a chain of title; they are searching for evidence pertaining to the location of the corners. Adjacent landowners often have plats or other documents that add details about the location of real property corners.

8.3. OFFICE PROCEDURE

Of all the questions commonly asked of the surveyor, "Where do you start?" ranks as number one. The surveyor is usually told the current land record identification of the particular parcel to be surveyed. That identification may be a lot number and subdivision name, an aliquot part of a section, a metes and bounds description, or just "the Jones's land." As in a Greek tragedy, which begins at the end before it returns to the beginning, the surveyor must trace back through the public records and his or her private files to the original tract from which the present parcel was created. This "paper trail" may require many hours of research before any fieldwork is started. This step is greatly shortened if the surveyor has previously surveyed other parcels of the same original tract. The value of prior knowledge in a particular area is a greater influence on the "regionalization" of private surveyors than any other factor.

The theoretical starting point for any survey is the same whether USPLS, metes and bounds, or platted subdivisions are involved. That theoretical starting point is the *record containing the creation of the original parcel*. In the USPLS, that starting point is the government surveyor's original notes. In the case of the metes and bounds states, that starting point is the original land grant. In the case of the platted subdivision, that starting point is the original subdivision plat. In cases where the platted subdivision was not monumented, recorded, or public use and acceptance of the created streets has not taken place, the starting point reverts to the parent tract.

8.4. FIELD PROCEDURE

Once the surveyor has gathered as much recorded and other written information as possible pertaining to the location of the real property corners, the field search for evidence of boundaries can begin. Each discovery is used to narrow the search for other corners. The most difficult corner to find is usually the first corner. After the first corner is found, the area of search is greatly narrowed for subsequent corners, facilitating their recovery. Potential artificial corner monumentation must be compared with the record for consistency before it is accepted. Among the questions that need to be answered are, "Who set the monument?", "Has it been disturbed?", "Is its location consistent with other monuments?", "Is it of the same material as called for in the plat?" and so on.

Suspected corner locations are examined for evidence of the monumentation of record. The advent of the electronic metal detector has been as great an advancement in the land surveying profession as the computer and the electronic distance measuring device. Large areas can be searched electronically for iron-based monumentation, and the area excavated can be greatly reduced. However, in the final analysis, the most important tool in the recovery of boundaries is the shovel. There are very few corner monuments that survive standing above the ground. The threat of earth moving equipment, brush cutting machinery, and other hazards—unintentional and otherwise—has led to the practice of burying boundary monumentation in order to insure that the location of that monument is undisturbed.

The surveyor may establish a control traverse first (section 6.9), and/or measure from known corners or accessories to isolate the search areas, compute theoretical positions for the corners, and then intensify the search in those areas. Recovery of corners that were originally documented by state plane coordinates is faster and more reliable when the control traverse used in the recovery attempt is also based upon the state plane projection system. There is no secret formula for boundary recovery, and the choice of measurement procedure made by the surveyor is in many ways incidental to the success of the boundary recovery. A successful recovery is one that finds the natural or artificial monuments at the corners of a real property parcel.

8.5. RENDERING A DECISION

Rarely is every corner monument found in any boundary survey. The ravages of time and the activity of agriculture or land development frequently destroy many of the artificial monuments placed by the original surveyor. This does not mean that the location of the corner is lost. The recovery of accessories to the corner reported by the original surveyor is a recovery of the monumentation for that corner, even if the artificial monument at the corner is gone. Often the relationship of buildings, tree or fence lines, or other terrain features to the corner being sought were reported on previous surveys for the subject or adjacent tracts.

Certainty of recovery is directly proportional to the relative relationship of reported terrain features to the corner being sought. Recovery of the original and undisturbed monument set at the corner is absolute. Other features, such as witness trees or buildings that are a known distance and direction from the corner can be used to recover a corner. Features that are nearest a corner, geographically and chronologically, are more likely to lead to a correct corner recovery than those that are remote from the corner being sought. A known corner 1 mile away is not nearly as reliable as a witness tree 10 feet from a corner. A survey plat claiming to have recovered an original monument produced 1 year after that original corner was set is usually more reliable than one produced 100 years later.

The time and effort spent in searching for physical evidence of the location of a corner is the dominant factor. The greatest source of dispute among land surveyors about the location of real property corners can usually be traced to a failure to discover all of the physical evidence relevant to a corner location. The tendency to simply ''lay out'' the deed dimensions of a parcel or to resort to mathematical computations that theoretically set the location of a corner often causes much confusion when, at a later date, additional physical evidence reveals the true location of the real property boundaries. A land surveyor's professional ability is a function of his or her thoroughness in the collection of physical evidence more than any other single factor. Excellence in mathematics, beautiful plats, sophisticated electronics, or professional appearance are meaningless unless a land surveyor's boundary recoveries are based upon all of the pertinent physical evidence that can be found.

Rendering a decision on the recovery of a property corner is an art. Formal education, continuing education, and extensive study will shorten the time it takes to excel in the art of corner recovery, but only experience in a particular geographical region will develop the skill needed for consistent and certain corner recovery. It is both proper and wise to request that a land surveyor inform you of the evidence and reasoning that led to the surveyor's decision on the location of recovered corners.

8.6. USPLS BOUNDARY RECOVERY

Any survey of a USPLS section or *any aliquot part thereof* begins with the recovery of section corners and quarter corners. The best record of where any corner was set is the original notes of the government surveyor. Plate 2 is a photocopy of facing pages from a typical government surveyor's notebook.

Although most government surveyors' notes are arranged in this form, some notes may not be. The first column lists the direction of the line being "run." This line was typically a magnetic observation that was "corrected" for the local magnetic deviation. The "correction" involved marking on the ground a "true north" (actually astronomic north) line by observing the star Polaris at the headquarters of the government surveyor. The reading of a compass, usually very large and considered more "precise" than the surveyor's field compass, was then compared to the "true north" line, and the deviation of the needle from the "true north" line was recorded. This is why the deviation or declination recorded in many government surveyors' notes are reported to the nearest minute when the compass used in the actual work could only be read to the nearest quarter degree. Of course, the survey work was usually many miles away from where the magnetic "deviation" was measured.

The second column lists the distances along that line that terrain features or other items of note that were encountered as measured from the beginning of the land lines. These terrain features encountered along the survey route are known as "passing calls." Passing calls are frequently very useful in

64

Pierre Robin Lacotte Rz R. 7 continued

N. 27½ E.	204.	02	to bayou Bienvenu 90 links wide
	232.	72	Set post to N. E. cor. of claim in prairie. —
			Then causing a flag to be set on the N. W. cor. of claim ran to it by
S. 83½ W.	17.	29	being Lacotte's backline
			Surveyed Nov. 2d 1872

Norbert Boudousquier O.B. 392

Sec. 4. T. 13 R. 12
Sec. 37. T. 13 R. 13
Sec. 44. T. 12 R. 13

Va. 7° 29' East

			From cor. 5/4 already reported
N. 26½ E.	16.	81	to intersection E. Bdy T. 13 R. 12 7.15 South of Township corner
			Set post, no trees
	24.	81	intersection of N. Bdy T. 13 R. 13 3.53 East of Township corner
			set post whereof bears
			live oak 4" S. 78¾ W. 10½ lks
			ash 5" N. 12½ W. 30½ "
	42.	30	to canal at its junction with Delaronde canal ∥ to this line
	43.	82	× canal at edge of prairie
	68.	53	to branch of bayou Mercier 30 lks
	104.	18	to other branch of " " 32 "

65

Norbert Boudousquier O.B. 392 continued

N. 26½ E.	116.	36	to N. W. cor. of claim, post in prairie —
			From lower front corner of Sec. 5 along old road in sugar cane
East	19.	36	Set lower front corner post in field. —
			Thence on lower line
N. 26½ E.	45.	80	leave field to weeds
	75.	02	to canal right angle to line
	77.	50	enter willow swamp
	89.	93	intersection of East Bdry. T. 13 R. 12 34.02 north of 1st mile post
			Set post cypress 4" × 4", no trees
	98.	95	to canal × protection levee
	99.	30	enter deep swamp
	141.	25	intersection of North Boundary T. 13 R. 13 at 22.92 East of Township corner, set post whereof
			cypress 4" N. 54° E. 21 lks.
			" 35" N. 87½ E. 153 "
	147.	00	leave swampy woods to boggy prairie of high grass, flags &c.
	188.	43	to branch of bayou Mercier 35 lks
	212.	98	to other branch of same bayou 34 "
	232.	72	to N. E. cor. of claim
			Sent flag to N. W. cor. already reported & ran to it on backline of claim
N. 89½ W.	19.	55	/ —
			Surveyed Nov. 5th 1872

PLATE 2

recovering corners set by the government surveyor, for they provide evidence of the location of the corner that is much closer than all other evidence, except for witness monuments (trees). If, for example, the notes indicated that at 38 chains from a found section corner the surveyor crossed a small creek (in survey jargon:. "a creek is called for at 38 chains") and set a quarter corner 40 chains, the first place to begin the search for the quarter corner is 2 chains past the creek, not 40 chains from the found corner.

The remainder of the page is reserved for a narrative describing and/or identifying the beginning of the line, the features encountered or "calls," the end of the line, the corners set, and the *accessories to each corner.*

8.6.1. Accessories to a Corner

Accessories to a corner consist of physical objects near the corner to which distances and directions are noted for the purpose of future restoration of the corner. Accessories were most often natural objects, such as trees, and were sometimes referred to as "witness marks" or "bearing marks." Accessories are as much a part of the corner monumentation as the corner post that may have been set by the government surveyor.

Witness trees are the most common form of accessory to a corner. These trees were chosen for their proximity to the mark, usually less than one chain (66 feet) away, as well as their size and species. In order to distinguish the accessory from other natural objects, the surveyor would scribe on the object specific letters or numbers. Ideally, three or more accessories accompanied each corner and were chosen so that they encircled the corner. In the narrative portion of the notes, the surveyor lists the distance and direction *from* the corner *to* the accessory. Because the accessories are a part of the corner monumentation, a recovery of the accessories is a recovery of the original corner monumentation.

8.6.2. Recovered Corners

Corners that are reset by the use of accessories or by the discovery of the original, *undisturbed* monument are called "recovered corners." Corners reset by use of the passing calls should not be considered as "recovered," for passing calls rarely were in close proximity to the corner.

8.6.3. Deciphering GLO Notes

Line 10 of Plate 2 (page 116) is the beginning of a line of a private claim. The line begins at the corner common to sections 1, 4, and 5. This is clearly an irregular section. Note that the variation of the compass reported by the surveyor is 7 degrees *29* minutes east, yet the directions observed for the claim boundary line are reported in increments of one-quarter of a degree. This inconsistency of precision may indicate that the instrument and personnel used to measure the magnetic variation were not the same as those used to measure this particular line. Later recovery of the line (reported as bearing north 26½ degrees east) showed the actual geodetic bearing at this point to be north 25 degrees 32 minutes 14 seconds east (plus or minus 5 seconds). This difference of only 1 degree between the 1872 magnetically based observation and the modern (1985) geodetic direction for the same line is well within the range of reasonably expected values.

According to these notes, the line intersected the northern boundary of the township at a point 24.81 chains along the line and 3.53 chains east of the township corner. A post was set to monument this intersection, and two bearing trees were reported. The first bearing tree is a 4-inch diameter live oak, 10½ links away at a bearing of south 78 ¾ west. The second is a 5-inch diameter ash, 30½ links away at a bearing of north 12½ degrees west. The post and the trees are the monumentation of this particular point in the form of corner monumentation and accessories.

8.6.4. Passing Calls

At 42.30 chains along this line, the surveyor encountered a canal "at its intersection with the Delaronde canal." This report of features encountered along a line are referred to as "passing calls." This particular passing call is important, because both canals may remain long after the post and trees that monument the last point are gone. Passing calls, when contradictory to corner monumentation, ought to yield to the corner and its accessories as proof of the location of a corner. This is only logical. Witnesses to an act who are nearest the action will carry more credibility than someone who is far away.

The art of recovering witness trees is highly refined in certain regions. Even a pattern of three or more stump holes that match the bearing and distances of the original witness trees is considered strong, if not conclusive, evidence of corner recovery. Unfortunately, in many areas, the activities of nature and human beings have erased all evidence of an original corner. This does not necessarily mean that the corner is "lost." Although the surveyor's notes recount the features that were geographically near the corner, other notes, survey plats, and events in the chain of title contain information that may be chronologically near the setting of the corner.

8.6.5. Perpetuated Corner

A section corner may be perpetuated through a chain of events leading up to features that remain at the time of a recovery attempt. For example, let us assume that a section corner and accessories are established in 1835. In 1898, a local surveyor, failing to find the original post, recovers the accessories and remonuments the section corner, according to the GLO notes, with an iron bar. In 1925, a local survey plat is made that notes the iron bar and several brick buildings, and shows several dimensions that relate the location of the buildings to the iron bar. In 1988, all that remains are the foundations of several of the buildings. These buildings have become, through perpetuation, accessories that monument the location of the original corner.

Corners that are reset by the discovery of a documented chain of evidence linking existing monumentation to the original monumentation by the government surveyor are called "perpetuated corners."

8.6.6. Corner by Common Report

It may be that the trail of evidence leading from the government surveyor's monumentation to the modern monumentation of a section corner is incomplete. A monument may have been accepted by an entire community as an original section corner without the documentation necessary to justify labeling it as a perpetuated corner. This acceptance by the community may be so complete that all of the aliquot parts created in each of the contiguous sections hinge upon that particular monument. If any alternate location for the corner were to be asserted, the whole community would be thrown into confusion. It is for this reason that, in the absence of better evidence, such monumentation may be accepted as marking the corner. This is known as a "corner by common report."

8.6.7. Corner Reset by Best Evidence

The most common type of corner or boundary recovery made by the land surveyor is to decide upon a location of a corner by the best available evidence. Passing calls, the relationship to other corners, evidence to the location of the corner not found in the government survey notes, and other sources of evidence often are sufficient to reset a monument at a corner location with certainty that the remonumentation is at the same location as the original corner.

8.6.8. Corner Reset by Proportionate Measurement

Finally, if absolutely no creditable evidence remains of the original location of a section corner, certain USPLS rules can be applied to "reset" the missing corner. Unfortunately, these rules, known as "restoration by proportionate measurement," are widely published by the U.S. government as well as possibly every book on surveying ever published, except this one. The word "unfortunately" is used, because the promotion of the rules governing the restoration of lost corners has encouraged many surveyors to resort improperly to proportionate measurement long before they have exhausted all of the possible sources of evidence.[1]

The recovery of the original section and quarter corners may not constitute a recovery of the boundaries or even assist in that recovery if actions took place, without regard to the sections of their aliquot parts, that re-established the property boundaries. If a boundary agreement or subdivision of the land took place without reliance on the original section corners, or if a boundary was established by adjudication, then the section lines remain but *cease to be boundary lines*.

8.7. METES AND BOUNDS BOUNDARY RECOVERY

The recovery of property boundary lines under the metes and bounds system of land identification varies from that of the USPLS in a few subtle but important ways. Although section corners in the USPLS can never move, the boundary corners of the metes and bounds system are subject to change. Because the monuments involved mark only boundary corners of contiguous lands and do not have the dual purpose of serving as control for other parcels, the adjoining owners are free to move or re-establish their common boundary lines. Boundary agreements, both formal and implied, as well as adjudication, can and will define new locations for property corners.

The paper trail from the present to the original parcel will still be present in the metes and bounds system, but the formal notes, regimented monumentation, and consistent use of accessories at each boundary corner are often absent. This places more reliance upon old survey plats and descriptions than does the USPLS recovery.

The parcels created under the metes and bounds system are not aliquot parts of the parent tract. The chronology of the creation and selling of divisions of a parcel become very important. The intentions of the parties to a division and sale of land parcels is not based upon a regimented procedure; rather, each division is a unique, individual act. Often it is quite difficult to decipher the true intent of the parties. These, as well as other complications, make the recovery of boundaries in the metes and bounds system a true art form.

The same rules of evidence apply as in any boundary recovery. Generally

[1] Division of Cadastral Survey, Bureau of Land Management, U.S. Department of the Interior, Stock Number 024-011-00012-7 *Restoration of Lost or Obliterated Corners and Subdivision of Sections—A Guide for Surveyors (1979)*.

speaking, the closer an object is, chronologically and geographically, to an original boundary corner, the more creditable that object is as a source of evidence to the location of the original corner. Old maps, plats, descriptions, and accounts, both recorded and unrecorded, may produce evidence pertinent to the recovery of a corner. Corners may be recovered, reset by best evidence, accepted by common report, perpetuated, or reset by proportional measurement, just as in USPLS corner recovery.

8.8. PLATTED SUBDIVISION BOUNDARY RECOVERY

Modern platted subdivisions present the simplest cases of boundary recovery. The requirements of most municipalities or local governments for the development and recordation of subdivisions produce well-monumented, clearly defined parcels with typically slight discrepancies between recorded dimensions and measured dimensions. Subdivisions that are well settled and well monumented, and that have clearly delineated boundaries rarely produce boundary conflicts.

Older subdivisions or subdivisions that were not settled and occupied may harbor discrepancies that will become evident only after an attempt has been made to posses several lots of the subdivision. The recovery of boundaries in subdivisions must begin with the limits of the subdivision block or lot involved. This recovery typically, but not always, involves public street rights-of-way. How much of the block is under physical possession (fences, houses, roadways, etc.) is of great importance to the recovery of lot boundaries. Rarely will any surveyor venture beyond block corners to verify boundary locations if the possession lines are in agreement with the deed dimensions.

The metes and bounds description that accompanies many deeds to lots in platted subdivisions will be considered by most surveyors as a supplement to the recorded plat. Discrepancies between the recorded plat and the metes and bounds supplemental description generally will be settled in favor of the platted data, unless it can be shown to be geometrically absurd. Just as with all boundary recovery, metes and bounds corners can be recovered by finding the monument set by the original surveyor, perpetuated, reset by best evidence, accepted by common report, or reset by proportionate measurement.

CHAPTER 9

EVALUATING SURVEY PLATS

Most of the preceding chapters have dealt, in a broad way, with the history and methods of boundary control and boundary recovery. Boundary recovery is the responsibility of the professional surveyor. The general information given so far will enable the reader to examine, evaluate, and use survey plats better. As in any profession, the ability, training, and competence of individual surveyors will vary. Even the best and most competent surveyors will occasionally make a mistake, miss a judgment call, or simply overlook something. Never assume that the surveyor is always correct. Each survey plat should be examined carefully for discrepancies or inconsistencies. The following checklist is designed to assist the reader in such an evaluation. Each of the items in the list will be explained in detail.

9.1. PLAT EVALUATION CHECKLIST

1. Determine the land record system used.
2. Determine the map projection used.
3. Evaluate the age of the survey.
4. Determine the purpose of the survey.
5. Examine the survey plat for gross discrepancies and completeness.

    a. North arrow and bearing base.

    b. Legal description.

    c. Date of the survey.

    d. Name and address of the surveyor.

    e. Signature and seal of the land surveyor.

    f. Adjoining properties.

    g. Dimensions of all sides of the property.

    h. Bearings of all sides or angles at each corner.

    i. Name of the client or purpose of the survey.

    j. Certification.

    k. Limiting notes or phrases.

    l. Area

    m. Scale.

6. Compare the survey plat with the deed for consistency.
7. Examine the survey plat for indications of easements.
8. Examine the survey plat for indications of encroachments.
9. Determine the accuracy standards required.
10. Determine needs not covered in the plat.
11. Contact the surveyor to discuss any excessive discrepancies or additional requirements.

9.2. DETERMINING THE LAND RECORD SYSTEM USED

It is not enough to know if the parcel in question is located in a metes and bounds state or in a USPLS State in order to determine the land record system that was used to describe a particular parcel. A platted subdivision may exist in any state. A metes and bounds division of a tract of land may be disguised as a platted subdivision by the seller's use of ''lot numbers'' in the sales. A reference to a portion of a section may refer to something other than an aliquot division, according to USPLS rules. Both platted subdivisions and USPLS parcels may have a metes and bounds description included in the deed as a supplement to the legal description. The importance of the differences between the various systems is only realized when the actual dimensions and locations of real property boundaries of a particular parcel are substantially different from the information found in the deed. The procedure for rectifying such discrepancies will vary according to the land record system used. Lots created in a platted subdivision do not follow the same rules as lots that developed sequentially in a metes and bounds system.

The words "south half" in a USPLS area mean something quite different from those same words in a metes and bounds system.

9.2.1. Recognizing a Metes and Bounds Description

The key to recognizing a metes and bounds controlled description is the lack of direct reference to the control principles of either of the other systems. Even if a parcel is located in a USPLS section and the section number, township, range, and principal meridian are a part of the deed, the parcel may still be a metes and bounds division of a USPLS section. Similarly, the presence of a lot number and subdivision name does not necessarily eliminate the possibility that a metes and bounds system is the controlling land record system at work in a particular situation.

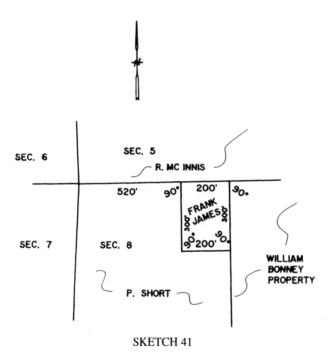

SKETCH 41

Sketch 41 is an example of a metes and bounds division of a USPLS regular Section. The Frank James property can only be described by a metes and bounds definition. This particular situation assumes that the line between sections 5 and 8 and the real property boundary line between P. Short and R. McInnis are the same. The line between P. Short and W. Bonney (720 feet east of the section corner) may be one-quarter of the distance to the quarter corner (a $\frac{1}{64}$ corner is ideally 660 feet east of the section corner). The description of this parcel may identify it as a portion of the NW$\frac{1}{4}$ of section 8, and so on. Even if the bounds called for are section lines for certain sides, the departure from the USPLS recognized method of division causes this to be a metes and bounds described parcel.

If the parties that created the parcel shown as the "Frank James" property intend for it to be bounded by R. McInnis, W. Bonney, and P. Short, it is important that the real property boundary lines be stressed in the metes and bounds description. If only the section line or the $\frac{1}{64}$he line is called for in the metes and bounds description, confusion may result if the section and $\frac{1}{64}$he lines are later found to be separate from the real property ownership boundaries.

The series of subdivision plats shown in Sketch 42 indicate that the original owner of the Smith estate sequentially created and sold portions of the property up to May 1, 1932. The deed to each of the lots contains a reference to the lot number, as well as a metes and bounds description. The fact that the May 1, 1932 survey plat is a recorded document does not cause the various titles created from June 3, 1911 to May 1, 1932 to be anything other than a metes and bounds controlled description. The titles for each lot were created at different times, and a precise chronology of that creation can be proven. In order for the division to be considered a platted subdivision, the May 1, 1932 plat must have been created and recorded *before* any of the lots were sold.

If, on the other hand, the June 3, 1911 plat had numbered that portion identified as the "Remainder of Smith Estate" as lot number 2, then this would constitute a platted subdivision. The significance of the difference would only be brought to light if the original Smith estate were not 900 feet wide, as the June 3, 1911 survey states it is. If the June 3, 1911 plat were a platted subdivision, then lot 1 and lot 2 would share proportionally the shortages (possession to the contrary excepted). If the original Smith estate were 850 feet wide, for instance, then the sequence of events portrayed in Sketch 42 would result in lot 4 being only 200 feet wide.

The description of lot 4 would call for the boundary on the east to be that of the western line of the P. Peters property and the boundary on the west to be the eastern line of lot 3. Because lot 3 existed long before lot 4 was created, then lot 3 must be granted all of the width called for, even though the remainder of the Smith estate was not 250 feet wide, as it was believed to be. These subtle nuances, interpreting intent, rely heavily on the system of identifying parcels. For this reason, it is quite important that the system in effect be properly identified.

9.2.2. Recognizing a USPLS

The use of the USPLS must include, as a minimum, the section number, the township number and quadrant, the range number and quadrant, and the

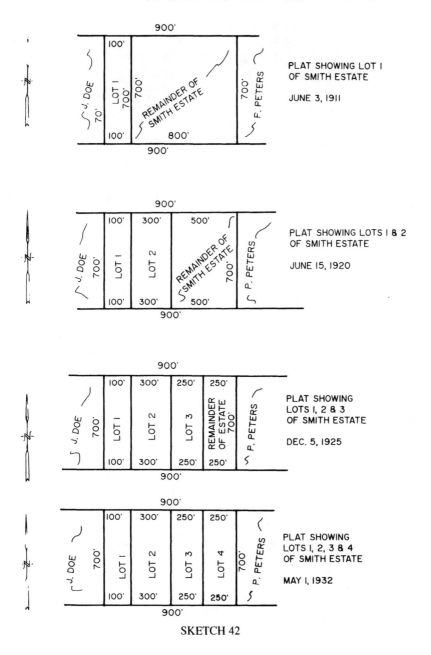

SKETCH 42

principal meridian. In some cases, the principal meridian may not be necessary if the state and county are reported *and* the state or county is served by only one principal meridian. Divisions of sections into smaller parcels must be USPLS lots, in the case of irregular or fractional sections, or must be created by the application of the rules of division governing the aliquot parts of a section.

Divisions of a USPLS section, regular or not, that do not conform to the USPLS regulations must be by platted subdivision or metes and bounds. Division of a USPLS lot is not provided for in the regulations. Sections, or their aliquot parts, must be divided by halves or quarters, based upon the *actual length of the sides,* not the theoretical length of the "ideal section." Special rules apply for irregular or fractional sections.

Divisions that are based upon the intention to transfer a specific area of land cannot be based upon the USPLS rules for the division of a section. A call for one-half of a section is not a transfer based upon a specific area. Because sections are never square, the words "south half of the acreage," "south 320 acres," and "S$\frac{1}{2}$" all have different meanings. Only the call for the "S$\frac{1}{2}$" is a USPLS division.

9.2.3. Recognizing Platted Subdivisions

Naturally, before the rules governing platted subdivisions can apply, a platted subdivision must have been created by someone. It is not absolutely necessary that the platted subdivision be recorded in the public records for it to be valid. It is not even necessary that *all* of the regulatory steps required by the local governing authority be complied with for the platted subdivision to be valid. The determination of the validity of a platted subdivision is a matter of law, and discussion of it is beyond the scope of this book. All of the following discussions about a platted subdivision will assume validity.

In order for a parcel to exist under the platted subdivision system, that parcel must have a unique identifying label. This usually takes the form of a lot "number" or "letter." Lots may also be called "parcels," "plots," or "tracts," among other things. The distinguishing feature that denotes a platted subdivision is the creation of two are more parcels of specific size from one or more existing parcels *where the division is based upon a drawing of the parent tract.* The drawing may be the result of a boundary survey conducted at the time of the creation of the subdivision or simply lines drawn upon an old map.

Divisions that intend to transfer parcels based upon the location of distinctive terrain features are metes and bounds transfers, even if a plat is produced to assist in the transaction. Divisions that adhere to the requirements of the USPLS are not platted subdivisions, even if a survey and a plat are produced and recorded for the parties involved.

The use of a subdivision "name" is quite common, especially in instances in which a large number of lots are involved. The use of squares or blocks is also common but not necessary. The creation of new streets or public rights-of-way are also common features of the platted subdivision of a parcel of land. Modern platted subdivisions are usually easily identified because of the many requirements placed upon the subdivider by the local governing authority.

9.3. DETERMINING THE MAP PROJECTION USED

The map projection used will have significant relevance to the degree of precision that may be reasonably expected of the information found on a particular survey plat or map. Directions, for example, are exceptionally sensitive to the distortions inherent in the tangent plane projection—so much so that bearings reported on plats using this projection system cannot be reliably compared with plats of adjacent tracts or even other plats of the same property.

9.3.1. Recognizing a Projectionless System

All projection systems commonly used in survey plats require that the bearings shown be based upon *angles measured* at the intersection of land lines. Maps that report bearings of the boundary lines as independently observed magnetic or astronomic directions are rarely based upon any projection system at all. The most common example of a "projectionless" map is the U.S. government township plat. These plats are simply a graphic rendition of the divers government surveyor's notes. Each bearing of each line is an independently observed direction (usually magnetic and "corrected" to "true north").

These directions are not two-dimensional; therefore, the novice will easily be misled. Land lines are defined as "straight" between corners, but all of the intricacies described in Chapter 4 apply. Surveys that were performed by "running" the boundary lines with a compass are usually "projectionless" surveys. The distances and directions of such surveys are typically of the lowest order of precision. This is not to say that these survey plats are without value. It may be that the value of the land being surveyed or the purpose of the survey (e.g., cutting of timber) may not warrant the higher cost of more precise surveys.

9.3.2. Recognizing a Tangent Plane Projection System

Tangent plane projection plats are the most common form of survey plats. The earmark of this type of plat is the use of a general reference to a meridian. Notes on the plat that state that the directions are based upon "true north" or some other meridian are strong indications that the map is a tangent plane projection. Notes that state that the directions are based upon the reported bearing of a recovered line almost exclusively indicate tangent plane projections.

Plats of very small parcels, such as urban or suburban lots, are often tangent plane projections. In the case of a platted subdivision, the directions shown on the plats of individual lots are frequently based upon the same projection system as the subdivision plat. In these cases, the directions shown on each lot plat should be consistent with similar plats of other lots in that same subdivision.

9.3.3. Recognizing a State Plane Projection System

State plane projection plats almost always state on the face of the plat that the directions are "state plane" or "based upon grid north." The appearance of state plane coordinates (SPCs) in the form of "x" and "y" values indicates a state plane projection. Because SPCs are established by an act of the legislature of each state, the surveyor is usually compelled by that act to clearly indicate when a state plane projection is being used.

Changes in the state plane projection system for each state will take place between 1983 and the year 2000. These changes are very complex and are necessary for the full implementation of space technology in high-precision geodetic positioning. Prior to 1983, all state plane projections were based upon the North American Datum of 1927 (NAD 27). After 1983, a new datum was introduced by NGS to replace NAD 27. This new datum is known as the North American Datum of 1983 (NAD 83). During a transitional period, maps or plats may be based upon either NAD 27 or the new NAD 83. After the end of the transition period, only NAD 83 will be allowed. The implementation of NAD 83 is individually legislated in each state.

9.4. EVALUATING THE AGE OF THE SURVEY

Evaluating the age of the survey is a little more involved than simply reading the date of the survey plat. The "age" of the survey is also a reference to the methods employed by the surveyor who conducted the work. The precision of the work, and thereby the confidence that can be placed in the accuracy of the results, is directly related to the "age" of the survey. As has been shown, government land surveys are not very precise when compared with the general level of capabilities of the survey profession during the same period. This may also be true of certain surveyors when compared with their peers.

Even late twentieth-century work may have been performed using old methods because of some special circumstance. A boundary line "run" by magnetic compass in 1980 is not more precise than one run in 1910. A distance measured using rigidly controlled procedures and a calibrated steel measuring tape in 1920 could be more accurate than a poorly controlled chaining in 1970. A telephone conversation with the surveyor who performed the work is the best way to determine the "age" of the survey. If that is not possible, the general history of survey procedures in the region may be obtained from local surveyors.

Most lending agencies have strict rules regarding the acceptable amount of elapsed time between the survey and the mortgage loan. These agencies have learned, through experience, that changes in the possession and use of the land can be quite rapid. Buildings can be erected or expanded. Fences can announce the commencement of acquisitive possession. There are a host of activities or occurrences that can alter the use and enjoyment (and thereby the title) of real property.

9.5. DETERMINING THE PURPOSE OF THE SURVEY

The mere existence of a survey plat is an indication that a contract, verbal or written, was formed between a land surveyor and persons interested in some feature of the parcel involved. If you did not request the survey, the possibility that the surveyor performed the work you want is very small. The variety and degrees of complexity in land surveys are such that many of the particular items of interest to you, such as easements or flood hazard, were not even reviewed by the surveyor, because the person who requested the work was not concerned about those particular details.

If the survey plat being reviewed was not ordered for the specific purpose to which it is being applied, then the survey information on that plat must not be important to the users. If a specific reason for the performance of a survey exists, the surveyor must be completely aware of that purpose so that she or he may tailor the work to suit the needs given. Often, items that are of particular concern are not discoverable by a survey alone.

9.5.1. Land Title Surveys

Surveys that are part of a title examination for the purpose of a change in ownership of a parcel of land require coordination between the surveyor and the title attorney. These particular surveys are known as "land title surveys." To avoid confusion over the shared responsibilities in a land title survey, the American Land Title Association (ALTA) and the American Congress on Surveying and Mapping (ACSM) have jointly established and published the criteria for land title surveys. This publication is called "Minimum Standard Detail Requirements for ALTA/ACSM Land Title Surveys."[1]

9.6. EXAMINING THE SURVEY FOR GROSS DISCREPANCIES

The surveying profession must recruit from human beings. This severe limitation means that surveyors may, from time to time, omit vital information or miss obvious discrepancies on the face of a survey plat or within a report. The most difficult blunders to detect in the survey profession occur during the final printing or typing of a survey plat or report. Examine every survey plat with a critical eye. Take special joy in detecting spelling errors, misprints, obvious omissions, or the transposition of numbers or letters. Distances or bearings that don't seem possible deserve an immediate response from the surveyor. If the discrepancy is the result of a refinement or a correction of a former value, the surveyor will tell you. Most of the items listed in this category are self-explanatory.

9.7. NORTH ARROW

The north arrow shown on a survey plat serves a twofold purpose. Primarily, the north arrow serves the purpose of orienting the plat. The north arrow also serves as an indicator of the bearing base used by the surveyor in conducting his or her work. Changes in bearing base or the use of an imprecise meridian will be reflected in changes in the values presented for the directions of lines. These changes are not relocations of the lines, just an assignment of new values. The absence of a north arrow makes the development of a metes and bounds description quite difficult. Considered by most surveyors as an essential item on any survey plat, the absence of a north arrow is sometimes common in cases of surveys in very dense, urban areas. If the survey plat that you are using lacks a north arrow and you feel that it needs to be shown, call the surveyor and have it added.

9.8. LEGAL DESCRIPTION

Depending upon the land record system in use, the survey plat must show the information necessary to identify the parcel. This does not mean that the

[1] See Appendix A.

complete verbiage of a proper metes and bounds description should be shown on the face of a survey plat in a metes and bounds land records system. A general or abbreviated version of the description will suffice to identify the parcel. The words, figures, and lines drawn upon the survey plat itself will complete the identification. In platted subdivisions and USPLS records systems, the survey plat caption ought to be in the form of lot, square, subdivision name, county and state or aliquot part, section, township, range, principal meridian (may be omitted), county, and state.

Regulations that require that the legal description be shown on the face of a survey plat are referring to the land identification requirements of the system in use. Unfortunately, some have interpreted this requirement to mean that a complete metes and bounds description must be printed on the face of every survey plat. The absolute impracticality of this misinterpretation will be better demonstrated in the exercises in writing metes and bound descriptions found in Chapter 12.

9.9. DATE OF THE SURVEY

The date shown on the survey plat serves to "fix in time" the data represented on the plat. Just as a photograph will present images that are locked in time, a survey plat contains the three dimensions of physical space as well as the fourth dimension of time. As survey plats age, the data shown on the plat may become obsolete. Buildings, fences, waterways, and rights in property all may be altered by time. A photograph of a 2-year-old girl may not accurately portray that same person 30 years later. Changes over time are to be expected in land parcels as well.

9.10. NAME OF THE SURVEYOR

Most survey firms produce survey plats that are drawn on standard paper that has the border, company logo, address, and other information already printed on it. It is surprising how many times this can lead to misunderstanding or oversight. In every survey firm, the responsibility for the collection and interpretation of data rests with one person. That person must be a professional surveyor registered in the state where the land is located. That individual's name must be shown on the survey plat.

Throughout this book, the need for communication between the user of a survey plat and the surveyor who performed the work is stressed. This is only possible if the surveyor is identified and the means of contacting her or him are clearly presented. A map that does not identify an individual as the surveyor responsible for the drawing is not a survey plat.

9.11. SIGNATURE AND SEAL

The signature and seal have a special importance in land surveying. This formal act of "signing" and "sealing" is a warranty by the surveyor that the information presented is the result of investigations, research, and measurements faithfully conducted according to the standards and procedures of the surveying community existing at the time that the survey was conducted. This is a guarantee that the surveyor *believes* that the information is a reasonably accurate report of the facts. Deliberate misrepresentation, distortion, misleading information, deletion of relevant facts, or fraud are pledged not to be present. This is a commitment of faithful service made to all who may see the plat. It is not a promise of infallibility.

9.12. ADJOINING PROPERTIES

The identification of all of the adjoining properties is normally consistent with the land record system used to identify the subject parcel. Metes and bounds plats usually identify the owners of adjacent properties by name. USPLS-based plats often show the entire section, with emphasis on the aliquot part or government lot that is being surveyed. Platted subdivision lots may refer to only the lot numbers of the neighboring lots. Platted subdivisions that consist of squares or blocks, as well as lots, may only refer to the square or block boundaries instead of the lots immediately adjacent to the lot being surveyed. It is also possible that the adjoining lots may be identified by a combination of systems.

9.13. DIMENSIONS OF ALL SIDES

The dimensions of all of the boundary lines of the parcel surveyed should be shown. If the dimension of a particular side cannot be measured, then an estimate and an explanation for the lack of a measured dimension should be shown. The parameters of the dimensions should be consistent. If the measurements are reported in U.S. survey feet, tenths and hundredths, then all sides should be in U.S. survey feet, tenths and hundredths. If directions are reported by bearings, then the bearings of all sides should be shown.

An exception to this rule occurs in cases in which, for historical reasons or for clarity, dimensions that are reported in the deed record but are no longer in use are shown on the plat, along with the modern measured value. A frequent form of this exception is use of the notations "actual," meaning the dimension measured during the present survey, and "deed" or "title," meaning the recorded or previously reported dimension for that same line segment.

9.14. BEARING OR ANGLES

The bearings shown on a survey plat must also be consistent in both parameters (e.g., degrees, minutes, seconds) and meridian. A bearing shown to the nearest second, because the precision of the measurement implied in the first case is much less than that implied in the latter. For example, a bearing of north 25 degrees east must be shown as "north 25 degrees 00 minutes 00 seconds east" in order to be entirely consistent with a bearing shown elsewhere as "north 27 degrees 15 minutes 33 seconds west." Naturally, presenting bearings in degrees, minutes, and seconds in one place and grads, for instance, in another, is a gross violation of the "consistent parameter" rule.

It is also awkward to define directions of lines by a combination of angles and bearings. Clarity and consistency are best served when one method or another is consistently used throughout a survey plat. The most common exception to this general principle is the "implied direction." Often, particularly in small parcels, lines are parallel or angles are clearly 90 degrees. In these cases, omission of bearings or angles is frequently accepted as minor or unimportant. Plats of urban or suburban homesites are frequently restricted to "legal size" paper ($8\frac{1}{2}''$ × $14''$), thereby limiting the amount of information that can be reasonably shown in such a small space. The use of "implied direction" is one way in which much of the clutter can be eliminated from small drawings.

9.15 NAME OF CLIENT, PURPOSE OF SURVEY

The name of the person for whom the survey was performed or a particular statement expressing the purpose of the survey can be an important indication of the extent of the work performed. Only the person named on the survey has a reasonable right to expect the survey to report the information that he or she requested. A survey for a particular purpose can only be reasonably used for that purpose. It is foolish to assume that a survey performed at the request of someone else will contain all of the information you want to see. It is equally foolish to assume that a survey performed for one particular purpose will fulfill all of the requirements of an alternate need. For example, a survey that was performed to confirm compliance with zoning restrictions may show the dimensions of a building that are not adequate for computation of the area of the floor space of that building. Zoning restrictions ignore recessed doorways; floor space computations do not.

9.16. CERTIFICATION

Special attention must be paid to the certification that appears on a survey plat. The phrase ''certified correct'' that was so popular in the past has been largely dropped by the surveying profession. This phrase has been found to be much too general and subject to interpretations never dreamed of by surveyors. ''Certified correct'' means not only that all of the information shown is correct (How accurate must a dimension be to be ''correct''?) but also that *all* of the information that the reader desires is shown. This interpretation requires that the surveyor know, at the time a survey is conducted, the needs of some unknown, future user of the survey plat. This is an impossible task to fulfill; consequently, surveyors have begun to *explain,* in the certification statement, as well as in special notes, the extent of work performed and the limits of the warranty promised by that work.

9.17. LIMITING WORDS OR PHRASES

The most pronounced of the recent developments in the relationship between the professional surveyor and his or her client has been the increased use of words or phrases on survey plats that spell out what was and what was not performed by the surveyor in the development of that plat. This clarification of the duties performed by the surveyor has caused some protest from the users of survey plats who did not realize that the plats that they had been using all along actually had these limitations.

Surveyors are not title abstractors. Surveyors do not, in the normal course of business, search the deed record or chain of title for easements, restrictions, zoning, or other legal instruments that may affect a parcel of land. Those parties who had incorrectly believed that surveyors were abstracting titles were quite upset when they were informed that the discovery and interpretation of recorded information rests with the legal profession. The discovery of visible acts, on the ground, that encumber a property, whether recorded or not, is the responsibility of the land surveyor. The attorney evaluates the written or legal factors encompassed in the title to real property; the surveyor evaluates the corporeal factors encompassed in the land itself. The combination of these two services constitutes a complete record of a real property parcel. Statements on the survey that explain the extent of work performed by a surveyor are there to ensure that no one mistakes a survey plat for an abstract of the deed record.

If it is important to the purpose of a particular plat that this information be shown, the surveyor must be presented with such written records as the user desires to be shown. Limiting phrases can be removed if the information not normally searched for by the surveyor is provided. These phrases are as much an integral part of the survey plat as are the bearings and distances, yet it is amazing how many persons, who would never contemplate asking a surveyor to misrepresent a dimension, demand the removal of these important informational phrases.

9.18. AREA

The area of a parcel is based upon a computation that involves several sets and combinations of measured values. For this reason, the area called for in a deed or reported in a survey plat holds the lowest ranking in the hierarchy of calls. Sales of property, on the other hand, often define the value of a parcel of land in terms of the area. Ironically, the one dimension considered to be the least reliable is often depended upon to set the most important aspect of any parcel of land—its monetary value.

Many surveyors will omit the area of a parcel unless that information is specifically requested by the client. This omission is not critical in most cases of surveys of small, rectangular, urban, or suburban lots in which the computation of the area is obvious or the value of the property is not directly determined by the area. In cases of large or complicated parcels, however, the area is an important part of the information that is relevant to real property parcels.

9.19. SCALE

The scale shown on survey plats represents the relationship of the dimensions on the original plat to the dimensions of the survey. Most survey plats that are handled by the client are *copies or reproductions* of the original drawing. Almost all reproduction processes will result in *distortions* of the plat. Paper will stretch or shrink in reaction to changes in humidity, and many other factors can have a dramatic impact on the fidelity of the scale on any plat. It is for these reasons that scaling dimensions on a plat are strictly reserved for *estimating* values. If a dimension is important, one should *never* depend upon a scaled measurement of that dimension. For this reason, many surveyors deliberately do not report the scale of their plats. Surveyors are only too happy to compute or measure dimensions that are of interest to the user of a survey plat.

9.20. COMPARING THE SURVEY PLAT WITH THE DEED

There often are inconsistencies between deed information and survey data. These inconsistencies are, normally, the necessary result of the improved or updated information about a particular parcel of land. The surveyor must make several judgmental decisions during the interpretation of boundary evidence discovered during a survey. It is very common for surveys of small, modern, urban home or business sites to agree exactly in every way with the information shown in the deed. This is because short distances are involved, rigid subdivision codes have been applied, and frequent street rights-of-way prevent the accumulation of small measurement errors into detectable proportions.

It is more common, particularly in cases of less than modern work, for the information contained in a deed to be slightly different from the information shown on a survey plat. Corrections to distances are quite common and may be the result of improved measuring techniques, corner interpretation, superior recovery methods, or simply new evidence, among other things. Directions (bearings) are even more likely to vary from the deed information than are distances. Except in cases of state plane projections, the accumulation of inconsistencies found in adjacent tangent plane projection surveys or the even more pronounced variations of magnetic meridians cause the directions shown on many survey plats to be suitable for little else than computing the angles formed at boundary line intersections.

Variations between state plane directions reported on a deed and state plane directions recovered during a particular survey should reflect only accumulated measurement error. The amount of acceptable variation depends upon the level of precision needed.[2] Contemplated changes in the state plane projection datum may cause a change in the values associated with the state plane directions (bearings) of boundary lines. The survey plat should report the datum associated with the directions shown.

[2] See Appendix B.

9.21. EXAMINING THE SURVEY PLAT FOR EASEMENTS

Easements are a nonpossessory interest held by one party in the real property of another whereby the first party is accorded partial use of said property for a specific purpose.[3] These rights that others may have may flow from one of two sources. The easement may be the result of a written agreement, or the easement may be the result of use. Easements may take one of two forms: They may be obvious (apparent), or they may be hidden.

Easements that are obvious, with physical indications of their existence on the ground, will be discovered by the surveyor. Easements that are hidden may not be discovered by the surveyor. Hidden easements that are written and recorded will be discovered by *a title examination only*. Surveyors do not normally perform complete title examinations unless they are requested to do so by the client. Most surveyors are not in the title insurance business. If you want to have all of the easements or other statutory restrictions associated with a real property parcel shown on a survey plat, *you must inform the surveyor of their existence*.

Title insurance companies are responsible for the majority of survey requests in the United States. Surveys performed for title insurance companies will not duplicate the title research work of the insurer. The surveyor will report all apparent easements, some of which may not be discoverable by a title search, and the title insurance company will report all *recorded* easements, some of which may not be discoverable in the field.

Compare each survey plat with the title records for evidence of hidden, but recorded, easements. If a survey plat does not report an easement that is in the title record, it is wise to inform the surveyor so that the surveyor may show this information on the survey plat. *Never rely on a survey plat as the absolute and sole source of land title information*. The survey plat is a field report of physical limits and conditions of a real property parcel. Title examinations and opinions are the professional preview of attorneys, not surveyors.

9.22. EXAMINING THE SURVEY PLAT FOR ENCROACHMENTS

Encroachments are physical objects (obstructions) that invade upon the rights of another.[4] Encroachments are evaluated by the degree to which the free

---

[3] American Congress on Surveying and Mapping and the American Society of Civil Engineers, *Definitions of Surveying and Associated Terms,* 1978 (rev.).
[4] American Congress on Surveying and Mapping and the American Society of Civil Engineers, *Definitions of Surveying and Associated Terms,* 1978 (rev.).

use and enjoyment of a real property parcel is affected. Few have any problem with recognizing and evaluating the impact of the encroachment of buildings, walkways, roadways, and other structures normally contained entirely within the boundaries of a designated parcel. The acquisitive act of a building extending across a property line is very clear; this is not so with fences.

9.22.1. FENCES AND FENCE LINES

Fences and fence lines deserve particular discussion here because, unlike all of the other forms of structures mentioned, they are normally intended to be "on the boundary." Boundary land lines are not physical things and, as we have seen, are defined to have a length but not a width. Fences are very real things and do have width. It is quite impractical to expect a fence to be "on the boundary line" without some portion of the fence being across the boundary line. There is much debate about what is measured when the surveyor is measuring a fence location.

Sketch 43 is an example of what has been offered by some as a way of interpreting the term "fence line." This interpretation of "fence line" defines facing of the fence as that which constitutes the land lines. It is the barbed wire, or the chain link, or the 1 × 6 facing, not the supporting fence posts, that obstruct movement across fence lines. Therefore, these obstructions must form the "fence line." This logical theory has a very fundamental flaw.

Let us assume that a landowner wants to construct a redwood fence on the boundary line between himself and a hostile neighbor. If the definition demonstrated in Sketch 43 were to apply, then the owner would center the posts 4 inches from the boundary line (the thickness of the 1 × 6 facing plus the 2 × 4 stringers plus $\frac{1}{2}$ the diameter of the fence posts). Setting the posts and bolting on the stringers can be accomplished without crossing the boundary line. Nailing on the 1 × 6 facing, as well as regular maintenance, would require the owner of the fence to trespass onto the neighbor's land. If the proposed definition of "fence line" were to prevail, no one could construct or maintain a fence "on line" without the full and continual cooperation of the adjoiner.

If, on the other hand, we were to define the fence line as being the centerline of the supporting fence posts, then $\frac{1}{2}$ of all fence posts would be across the boundary line. There would be land between the boundary line and the facing of the fence that could only be maintained by the neighboring landowner. Complicating this situation is the fact that most fence posts are not perfectly vertical. Where is the fence line in these cases? At the ground? At the top of the fence? It is obvious that practicality demands toleration of

IDEAL WOOD FENCE SECTION

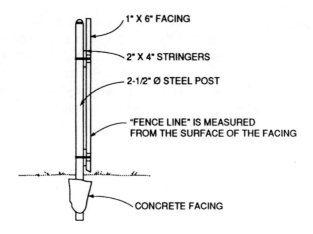

1" X 6" FACING

2" X 4" STRINGERS

2-1/2" Ø STEEL POST

"FENCE LINE" IS MEASURED
FROM THE SURFACE OF THE FACING

CONCRETE FACING

IDEAL BARBED WIRE SECTION

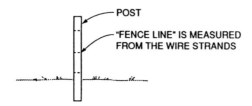

POST

"FENCE LINE" IS MEASURED
FROM THE WIRE STRANDS

IDEAL PAGE FENCE SECTION

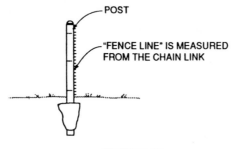

POST

"FENCE LINE" IS MEASURED
FROM THE CHAIN LINK

SKETCH 43

a certain amount of encroachment, as well as trespass, in order for a boundary fence to exist.

It is for these reasons of practicality that the rigid definition of the location of a "fence line" demonstrated in Sketch 43 is being replaced by one that defines a fence as being "on line" when any visible part of the fence is on line. Under this "liberal" definition of a fence line, fences may be constructed without continual trespass for maintenance and without the fear that a slight encroachment of facing or post could be interpreted as an acquisitive act. Each locality may differ in this interpretation, based upon regional court decisions.

9.23. DETERMINING THE ACCURACY STANDARD REQUIRED

The American Congress on Surveying and Mapping has developed a set of classes of surveys based upon expected land use. These general classifications have been mirrored, to varying degrees, by many state surveying organizations or regulatory boards. The ACSM goes on to list several procedural and closure requirements for each class. These requirements are somewhat technical in nature, only applicable during the measurement and computational phases of a survey, and very difficult to assess without field notes or working papers. The list that follows presents the class of survey, as defined by the ACSM publication, as well as this author's interpretation of what the *effect* of the recommended procedure and computational requirements would be. The ACSM definition is set within quotation marks.[5]

9.23.1 "CLASS A. Urban Surveys"

"Surveys of land lying within or adjoining a City or Town. This would also include the surveys of Commercial and Industrial Properties, Condominiums, Townhouses, Apartments, and/or other multi-unit developments, regardless of geographic location." Generally speaking, the actual relative positions of distinct features within a "Class A" survey may vary from the position published on the plat by a maximum distance obtained by the equation (Variation = 0.03 feet + D/15,000), where "D" is the distance between features. For example, if the distance between two boundary monuments is shown to be 1,347.56 feet on a survey plat, the acceptable variation would be about 0.12 feet (0.03 + 1,347.56/15,000). Therefore, the actual distance between monuments may be anywhere from 1,347.44 feet to 1,347.68

[5] See Appendix C.

feet and still be within the tolerances of the "Class A" category. Similarly, the bearings shown on the survey plat have an allowable variation of about 15 seconds (relative to the meridian or bearing base and depending on the length of the line). A published bearing of north 27 degrees 34 minutes 05 seconds east would be within tolerances if the actual bearing were anywhere between north 27 degrees 33 minutes 50 seconds east to north 27 degrees 34 minutes 20 seconds east. This, of course, does not apply when *different bearing bases are used.*

9.23.2. "CLASS B. Suburban Surveys"

"Surveys of lands lying outside urban areas. This land is used almost exclusively for single family residential use or residential subdivisions." The result of the application of the recommended tolerances for the procedures and computations of the "Class B" survey are positional accuracies or variations of about $0.05 + D/10,000$ feet and directional variations of about 20 seconds. The dimension of 1,347.56 feet given in the previous example would be considered within tolerances of a "Class B" survey if the actual value were anywhere between 1,347.38 feet and 1,347.74 feet. The bearing of north 27 degrees 34 minutes 05 seconds east would be considered within tolerances if the actual bearing (relative to the same base) were anywhere between north 27 degrees 33 minutes 45 seconds east and north 27 degrees 34 minutes 25 seconds east.

9.23.3. "CLASS C. Rural Surveys"

"Surveys of land such as farms and other undedeveloped land outside the suburban areas which may have potential for future development." The tolerances under "Class C" are about $(0.07 + D/7,500)$ for positions and about 30 seconds for directions. The acceptable range of values for our example dimension of 1,347.56 feet broadens to between 1,347.31 feet and 1,347.81 feet, and the directional range varies between north 27 degrees 33 minutes 35 seconds east and north 27 degrees 34 minutes 35 seconds east.

9.23.4. "CLASS D. Mountain and Marshland Surveys"

"Surveys of land which normally lie in remote areas with difficult terrain and usually have limited potential for development." The requirements for a "Class C" survey roughly translate into acceptable positional variations of $0.10 + D/5,000$ feet and acceptable directional variations of 40 seconds. The acceptable limits of our examples would then be 1,347.19 feet and

1,347.93 feet for position, and north 27 degrees 33 minutes 25 seconds east and north 27 degrees 34 minutes 45 seconds east for directions.

These example variations are not presented as absolute limits but merely as guidelines intended to give the reader a "feel" for the range of reported dimensions that one should reasonably expect for each "class" of survey. Users should evaluate their own tolerance for deviations based upon an assessment of the particulars of a given survey. Tolerances more restrictive than "Class A" are possible but prohibitively expensive. An old country surveyor's cliché best remembered when one is deciding upon the tolerances required is: "We're measurin' a fence line, not buildin' a violin."

9.24. DETERMINING NEEDS NOT COVERED IN A SURVEY PLAT

Every user of land survey data has his or her own particular needs, as dictated by the purpose of the survey. The list of items in category 5 of the ALTA/ACSM Minimum Standard Detail Requirements (Appendix A) represents the minimum requirements usually associated with the typical boundary survey. It is very possible that your particular needs will exceed the minimum requirements or be beyond the limits expressed by special notes or the certification.

Chief among the needs not commonly covered by the typical survey plat is the existence and location of all easements or servitudes affecting the parcel. Even in the absence of restrictive words or phrases, it is very wise to ensure the inclusion of recorded easements or servitudes by providing the surveyor with an abstract of the title or, at the very least, by comparing the survey plat with the deed and title information yourself. The surveyor will be delighted to add any information contained in the chain of title, even if it is not apparent on the ground.

9.25. CONTACTING THE SURVEYOR

It is vital that, before using the information found on any survey plat, the user of the information and the developer of that information have a clear and concise agreement on the limits of research and the purpose of the plat. This is not difficult if the survey plat was developed from a direct request by the user. Full communication, at the time the survey is requested, will provide this important link. If the user is not the same person as the one who is ordering the survey, full communication is not possible until the user contacts the surveyor directly.

The user should simply pick up the telephone and call the surveyor who developed the plat. The user should tell the surveyor how the plat is to be used and should ask if the surveyor anticipated that particular use. The user should question the surveyor about *anything* that appears to be odd or unusual. If the information is correct, the surveyor will happily explain it. If the information cannot be easily explained or is incorrect, the surveyor will gratefully revise the information and send an updated version of the plat or report. Surveyors want their work to be reviewed as carefully, as critically, and as often as possible *before* reliance is placed on the information presented. Often, the only critical review possible is in the surveyor's own office. When external critical review is possible, the surveyor welcomes it. The saddest words ever heard, usually after an expensive undertaking, are, "I thought that looked funny!"

Certifications or limiting words and phrases on survey plats that identify the plat as not providing all of the information that is desired do not render the plat useless. Often the desired information can be added or the limiting phrase reworded if the necessary information is provided to the surveyor. Plats that report that easement information is not shown, for example, may have that limiting phrase removed if the surveyor is provided with all recorded easement information. It is a great disservice to future users of a survey plat if any surveyor were to remove the limiting words from a survey plat without also eliminating the need for the phrase.

If the surveyor cannot be contacted, it is foolhardy to use any survey plat developed by that surveyor for anything other than a general graphic aid for the parties involved. The surveyors who are active in a particular area can provide insight on the abilities, typical procedure, and reasonable expectations of accuracy of deceased or retired surveyors. Often the files of inactive surveyors are sold or given to active local professionals for safekeeping and referral. The state or local professional surveyor's association can be contacted for assistance in locating such files.

CHAPTER 10

EXERCISES IN EVALUATING SURVEY PLATS

The following exercises are designed to clarify the recommended guidelines for evaluating survey plats. These guidelines are very general. Several states have developed, either through their state board of registration or through their professional societies, a minimum standard of practice. Such states may require information in addition to that shown as typical in these example plats. None of these example plats is presented as an ideal example of a survey plat. All of the examples are drawn from actual cases but are greatly simplified, and appropriate name and location changes have been made. The strengths and weaknesses of each example will be discussed so that the reader may ponder them and apply the thoughts expressed to real survey plats. The numbers and letters in parentheses ( ) refer to the checklists mentioned in Chapter 9. Each of the exercises will take the form of a scenario. The reader is not expected to agree completely with each and every step in the scenario. Indeed, part of the learning experience is reflecting on and pondering how differently the reader may have proceeded.

10.1. THE CASE OF THE THREE PARTNERS

Ms. Cazes has the deed to lot 2 of section 11, township 3 north, range 5 west, Walnut County, Nebraska. The deed simply refers to the property as "lot 2, T3S, R5W, Walnut County Nebraska." Along with the deed, which she inherited from her father, is the map shown in Sketch 44. Ms. Cazes's father was one of the original three partners who together owned section 11.

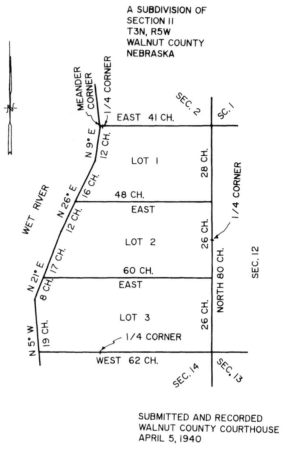

A SUBDIVISION OF
SECTION II
T3N, R5W
WALNUT COUNTY
NEBRASKA

SUBMITTED AND RECORDED
WALNUT COUNTY COURTHOUSE
APRIL 5, 1940

SURVEY BY B. P. WHITE, P.E.
APRIL I, 1940

SKETCH 44

The record indicates that the three caused the property to be subdivided and that each received one of the lots.

Ms. Cazes now wishes to sell lot 2. Evaluate the original plat and assess its usefulness in such a sale.

1. The presence of a section, township, or range might cause the reader to think that the controlling land record system here is the USPLS. This is only partly true. If lots 1, 2, and 3 were not formed by the government surveyor or by the application of the USPLS rules, then the lots were formed by a platted subdivision. The exterior boundaries of the subdivision, being the boundaries of section 11, fall un-

der the recovery rules of the USPLS. The lots do not. The land record system for the individual lots is the platted subdivision system.

2. The use of the cardinal directions, as well as the fact that all directions are reported to the nearest degree, suggest that the bearing of each line was independently observed. This is a projectionless map. The plat is simply a drawing of the directions and distances observed for each line.

3. The date indicates that the work was done in 1940. All distances are rounded off to the nearest chain (66 feet) and all directions, to the nearest degree. This indicates a crudeness of measurement that is more typical of 1840 than 1940.

4. The purpose of the survey is obviously to document a division of the section into three lots. The dimensions at the east boundary indicate a desire to divide the lots by near equal division of this side.

5 a. The north arrow is shown with no reference to a bearing base. The crudeness of the values suggests that a magnetic compass was used, and it is very unlikely that a resurvey of any of the lots will duplicate the directions reported on the subdivision plat.

   b. The legal description of any of the lots is "lot——— of the subdivision of section 11, T3N; R5W, Walnut County, Nebraska, recorded April 5, 1940" and is shown on the survey plat. The absence of a reference to a principal meridian will not be of any consequence if only one principal meridian serves all of Walnut County.

   c. The date of the survey is shown, along with the recordation date. This must be verified by examining the records.

   d. The name of the surveyor is shown, but not the address.

   e. The survey plat is not signed or sealed, but the age of the document indicates that it probably followed the accepted procedure for a subdivision of land at that time and, if recorded, will probably defeat any attempt to refute its authenticity.

   f. All adjoiners are shown by section numbers or as the Wet River. If the Wet River were navigable at the time of the original government survey, this would be a riparian boundary.

   g. All dimensions are shown and are unusually even. The dimension of 80 chains for the east boundary of the subdivision is particularly suspicious.

   h. The bearings of all sides are shown, and they, too, are suspiciously close to the directions typically called for in the government surveys.

    i. The name of the client is absent, but the purpose of the survey is obviously to document the creation of three lots.

    j. There is no certification statement.

    k. There are no limiting words or phrases. The drawing is one of such little detail that one could reasonably expect that buildings, fences, driveways, and other features exist but are not shown.

    l. The area is not shown.

    m. The scale of the drawing is not shown.

6. A comparison with the deed is consistent with the plat. The deed does not contain a metes and bounds supplementary description. The original government township plat of T3N, R5W, dated 1866, reports the exact same bearings and distances for the exterior of section 11 as does the subdivision plat.

7. There are no easements shown, but, considering the age of the plat, this is not surprising.

8. No improvements are shown at all. This may be an indication that the improvements were present but deemed unimportant or that the land may actually have been vacant.

9. The accuracy standards, as indicated by the dimensions shown, are so crude as to fall below any of the modern "classes" of surveys.

10. There is a wealth of information that is not addressed that would be of importance in a sale of this property. The deed record must be reviewed (title examination) for recorded easements, rights-of-way, possible partitions, and so on. Buildings, bounding fence lines, corner monumentation, and other physical features need to be examined. Clearly, this parcel must be resurveyed, and an abstract of the title record must be made prior to any sale. As with any purchase of real property, a title insurance policy is a must.

11. After attempts have been made to contact B. P. White, it is discovered that he is deceased.

10.1.1. Conclusion

The fact that the bearings and distances reported for section 11 by the government surveyor and those reported by B. P. White for the exterior of the subdivision of section 11 are exactly the same is conclusive proof that White did not resurvey or recover the boundaries of section 11 prior to the drafting of the subdivision plat. It is highly probable that White simply drafted the subdivision from the government plat and never even visited the property. This is a "paper survey" and serves no other purpose than to indicate the intended division of the section. Lot 2 must be resurveyed by recovering as many of the original section, quarter, and meander corners as possible.

A survey of lot 2, section 11, T3N, R5W, Walnut County, Nebraska was ordered, and Sketch 45 is a copy of the resulting survey plat.

By applying the same examination procedure as before, we can discover the following information.

1. The external boundaries of Section 11 are reported by J. Ford as being recovered. The distance between the NE corner of section 11 and the east ¼ corner is only 6.7 feet less than the original government survey's 40 chains (2,640 feet). The distance between the SE corner of section 11 and the east ¼ corner is 41.39 feet less than 40 chains. The line between the NE and SE corners of section 11 is not a straight line. All of these signs indicate a recovery of the east line of section 11 based upon a proper recovery of the corners.

2. The presence of an "x" and "y" at the NE corner of section 11, as well as the note "Grid Bearings NAD 27" indicate that the state plane projection system for Nebraska is being used. Ford's references could be elaborated in order to define better the projection system used. Nebraska law may require special wording or notes when state plane coordinates (SPC) are being used.

3. The survey plat indicates directions and distances that are defined to a second of an arch and to a one-hundredth of a foot. The date hints that a large portion of the data shown on the survey may have been previously collected by Ford during other surveys or that 1987 experienced an unusually mild January.

4. Because we requested the survey, the purpose is known. When the survey was requested, a full explanation of the purpose and expected use of the survey was made to Ford.

5. a. The north arrow and bearing base are shown.
    b. The legal description "lot 2 of section 11, T3N, R5W, 6th P.M. Walnut County, Nebraska" is shown on the survey plat.
    c. The plat is dated "January 18, 1987."
    d. The name and address of the surveyor is shown.
    e. The plat is signed and sealed.
    f. The adjoining properties are shown by lot number, natural feature, or by USPLS designation as applicable.
    g. All dimensions of all sides are shown. The north line of lot 2 shows a total length of 3,005 feet, *plus or minus*, meaning "more or less." This is appropriate because of the indefinite nature of

LOT 2 OF SEC. 11
T3N, R5W
6TH P.M.
WALNUT COUNTY
NEBRASKA

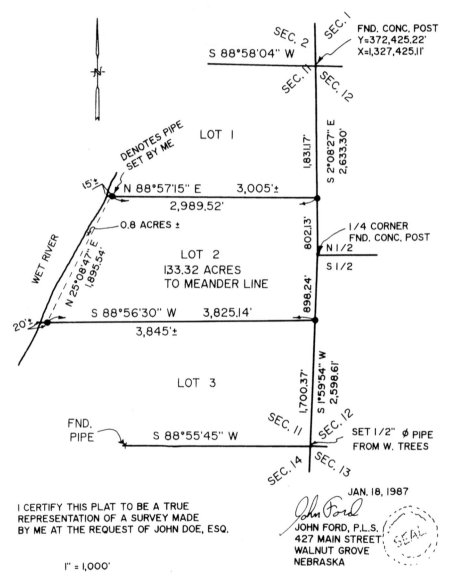

I CERTIFY THIS PLAT TO BE A TRUE
REPRESENTATION OF A SURVEY MADE
BY ME AT THE REQUEST OF JOHN DOE, ESQ.

JAN. 18, 1987

JOHN FORD, P.L.S.
427 MAIN STREET
WALNUT GROVE
NEBRASKA

1" = 1,000'

SKETCH 45

riverbanks. This distance is shown on the subdivision plat as 48 chains (3,168 feet). The distance between the NE corner of lot 2 and the NW meander corner of lot 2 is shown as 2,989.52 feet. The distance from the meander corner to the Wet River is shown as 15 feet, *plus or minus*. The sum of these two distances is 3,004.52 feet. This is consistent with the shown total of 3,005 feet, plus or minus. It would be scientifically incorrect to express the distance as "3,004.52 feet, plus or minus," because this would imply that the magnitude of the uncertainty was in fractions of a foot, whereas the plat clearly expresses the magnitude of uncertainty in feet (15 feet, plus or minus). Similarly, 20 feet, plus or minus, added to 3,825.14 feet is consistent with the total 3,845 feet, plus or minus. The subdivision plat reports this second distance as 62 chains (4,092 feet).

  h. The bearings of all lines, including the new meander line (north 25 degrees 08 minutes 47 seconds east), are shown. The fact that the bearings are expressed to the nearest second suggests that a high degree of precision may be accorded to these values.

  i. The name of the client is shown. The purpose of the survey is not stated.

  j. The certification states that the information shown on the survey plat accurately reflects the results of the survey performed by Ford. It does not claim that the survey was performed to any particular standard, nor does it warrant any of the information as "correct."

  k. There are no limiting words or phrases on the face of the plat.

  l. The area is shown to be 133.32 acres to the meander line. This means that the strip of land between the meander line and the Wet River is not included in the acreage figure shown. This meander line is a new line established by Ford and is not to be confused with a boundary line or the original meander line by the government surveyor.

  m. The scale is shown as "1″ = 1,000′." This may be checked by measuring a few dimensions with a ruler. Paper may stretch in one direction more than another, so any check should include east–west as well as north–south distances.

6. The deed, only referring to a lot number, is consistent with the survey plat. The original subdivision plat is not entirely consistent with the survey plat. In most cases, the deed will refer to a subdivision plat, making that plat a part of the deed. The failure of the deed to refer to the subdivision plat, in this case, is not of any consequence.

The discrepancies between the subdivision plat and the survey plat of lot 2 should be explained by the surveyor.

7. There are no easements shown. There is no note by the surveyor indicating whether a search for recorded easements was made by his office. This needs clarification.

8. There are no fence lines, buildings, or other structures shown, and there is no note indicating that the property is vacant. This needs to be clarified.

9. The presence of SPCs, distances specified to the one-hundredth of a foot, and bearings specified to the second indicate a high order of precision. This needs to be verified.

10. A review of the title abstract does not indicate any recorded easements, servitudes, rights-of-way, or any other encumbrances to the property. The absence of fences should be explained, as well as the lack of buildings. Perhaps a statement to the effect that these things do not exist is appropriate.

11. The surveyor can now be contacted to respond to the list of questions that have been raised. From this conversation, we learn that the section and quarter corners all have a long history and are well recognized, through perpetuation or recovery of accessories, as the original corners. The work was performed and controlled under the state plane projection system for Nebraska (south zone), North America Datum of 1927. Some of the data was developed over several years of fieldwork in the adjoining sections, although the momenutation of the north and south boundary lines of lot 2 and the Wet River was performed in response to the survey request. All work conformed to the requirements of a "Class A" survey.

Ford has affirmed that the survey plat was developed with the intent that it be used in a transfer of title. Ford also has reported that he believes that the original subdivision plat was an "office" survey that simply divided, on paper, the government version of the dimensions of section 11. His dimensions and directions for lot 2 are his professional opinion of the actual boundaries of lot 2, based upon his interpretation of the intent of the subdivider and the constraints of the true configuration of section 11. It is not known if the Wet River shifted east or if the original subdivision plat incorrectly reported the river's location. Ford agreed to add to the plat, in parentheses, old dimensions and directions shown on the subdivision plat to clarify that the existence of the discrepancies is known.

Ford reported that he did not perform a title search for recorded easements or other title restrictions. The property is vacant, wooded lowlands, and subject to inundation. The construction of a reservoir, upstream on the Wet River, is underway and may change the condition of the land.

We have had all our questions about the land answered. A modern survey plat and close communication with the land surveyor who developed the plat have made it possible to sell the parcel with a better understanding of just what was being exchanged.

There are some who may balk at accepting a survey of lot 2 that is, apparently, at such variance with the subdivision plat. Those so concerned would do well to reflect on the fact that, if the original subdivision plat version of lot 2 were simply copied, there would be an overlap of 163–247 feet on the east boundary and over 16 feet on both the north and south boundaries. The buyer would be paying for 10.5 acres of land that does not exist or is, at best, under contention. A modern survey of a parcel of land should be ordered only if one really wants to know the truth.

Surveyors are like doctors or lawyers in this respect: If the surveyor's professional opinion and explanation of that opinion are not satisfactory, then a second opinion should be sought. The very worst thing that one can do is to attempt to persuade the surveyor to distort the survey plat to match the deed. Land cannot be created by words upon paper. A land surveyor is consulted because of his or her special knowledge of measurements and boundary recovery. Don't reject the surveyor's opinion just because it "rocks the boat," but don't blindly accept that which cannot be explained to your satisfaction either.

10.2. THE LAND GRABBER

Let us assume that you intend to purchase a parcel of land on New York State highway number 10, in Dutch County, near the town of Dutchville, N.Y. The owner bought the property in 1978, at which time it was surveyed by the owner's cousin (now deceased), who was a surveyor from Albany. The papers presented to you, by the owner, describe the property as follows:

> . . . That portion or parcel of land bounded on the West by the land of Robert Beckmann, on the North by the lands of Gilbert Hirt, on the East by the lands of John Peters as defined by the old Cromwell Grant line, and on the South by New York State Highway #10 and containing approximately 5 and one half acres commencing and point of beginning at the old stone borne located on the Cromwell Grant line at the North side of New York State Highway Number 10; thence, S 25 E 50' to the centerline of Highway 10; thence, S 65 W 350'; thence, N 25 W 50'; thence N 25 W 700'; thence, N 65 E 350'; thence, S 25 E 700' to the point of beginning. . . .

This is a poor description, at best, but it is all that the owner has. The local surveyor is Mr. J. D. Times, who is highly recommended by the owner as being "the only man for the job."

In your initial conversation with Times, you advise him that you intend to purchase the property for investment purposes and that you are in the process of obtaining title insurance. Times requests that, if you send him copies of "all title papers and previous surveys" in your possession, as well as a copy of the results of the title examination by your title insurance company, he will perform the survey. The title insurance company representative had previously stated that the survey could be done and the plat forwarded to while the title examination was underway, and you mention this to Times. The sale has a strict time limit, so you urge Times to start immediately. Times agrees to start as soon as a copy of the deed is delivered.

The survey is completed, and several copies of the survey plat are delivered to you. Retaining one copy of the survey plat, you forward the rest to the title insurance company. Sketch 46 shows the survey plat received by you.

While the title insurance company is completing its work, you decide to evaluate the survey plat, using the checklist that appeared in Chapter 9.

1. The land record system is a metes and bounds system.
2. There is no clear indication of whether a map projection has been used (meaning angles were measured) or whether the plat is simply the result of observed bearings.
3. The survey appears to be the result of recent work. The bearings seem to be imprecise, because the minutes and seconds of direction are not shown. This may not be vitally important, considering the small size of the parcel. The distances are all reported to the nearest foot. This may be the result of imprecise work, or it may simply be that Times did not show the tenths and hundredths of a foot if those figures were zero.
4. The purpose of the survey was discussed in depth between you and Times.
5. a. A north arrow is shown, but the bearing base is not defined.
   b. The caption of the plat describes the parcel as "The Lands of A. A. Anderson, Dutch County, New York." This may not qualify as a "legal description," depending upon the cadastral records of Dutch County.
   c. The date of the survey is shown, and the work appears to be modern, except for the lack of precision reported in the bearings and distances.

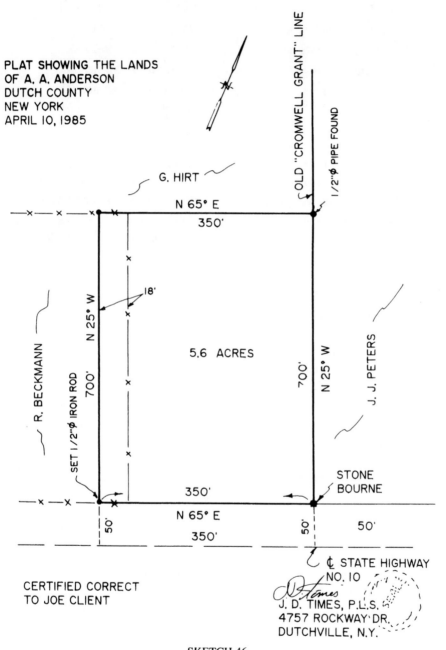

PLAT SHOWING THE LANDS
OF A. A. ANDERSON
DUTCH COUNTY
NEW YORK
APRIL 10, 1985

G. HIRT

OLD "CROMWELL GRANT" LINE

1/2"∅ PIPE FOUND

N 65° E
350'

N 25° W

18'

5.6 ACRES

R. BECKMANN

SET 1 1/2"∅ IRON ROD

700'

700'   N 25° W

J. J. PETERS

STONE
BOURNE

350'
N 65° E
350'

50'   50'   50'

₵ STATE HIGHWAY
NO. 10

CERTIFIED CORRECT
TO JOE CLIENT

J. D. TIMES, P.L.S.
4757 ROCKWAY DR.
DUTCHVILLE, N.Y.

SKETCH 46

    d. The name and address of the surveyor are shown.

    e. The signature and seal of the surveyor are shown.

    f. All adjoiners are identified.

    g. All dimensions are shown and appear to be consistent.

    h. All bearings are shown. The bearing base or method of determining the bearings should be clarified.

    i. The name of the client is shown.

    j. The plat is "certified correct" to you. This particular wording of the certification is extremely broad and is becoming very unpopular among surveyors. Your understanding of just what was done by Times would be improved by a more precise certification.

    k. There are no limiting words or phrases, in spite of the fact that there is an apparent encroachment of some 18 feet along the western boundary line.

    l. The area is shown as 5.6 acres, but it is not explained whether that includes the encroachment or excludes it.

    m. There is no scale indicated.

6. The directions and distances are exactly the same as the deed description furnished to Times. As in the old "B" westerns when things were "too quiet," this might be a case of things being "too good," but, considering the small size of the parcel, it is quite probable that a match of distances exists within the acceptable range of precision.

7. There are no indications of any easements, nor is there any note about the extent to which easements were searched for. The conversation with Times when the survey was ordered clearly indicated that Times would not perform a title examination.

8. There is an extensive fence encroachment along the entire west boundary line with R. Beckmann. This is a serious encroachment that encumbers $\frac{1}{3}$ of an acre.

9. The property is quite valuable and subject to commercial development. It is not possible to tell from the survey plat what the accuracy standards were for this survey.

10. The title records indicate a 10-foot easement "across the entire rear or northerly line" of the parcel. This information should be added to the plat. Clarification is required on the encroachment. Times needs to investigate, if he hasn't already done so, the titles to the Beckmann and the Hirt properties.

11. You contact Times with a list of questions resulting from the exam-

ination of the survey plat and receive the following clarifications. The survey was performed by "side shots" from a random traverse around the parcel. The bearings shown are based upon the reported direction of the "Cromwell Grant" line, as monumented by the stone borne and $\frac{1}{2}$-inch-diameter iron pipe (therefore, the plat is a tangent plane projection). Times estimates that the distances shown are accurate to plus or minus $\frac{1}{2}$ inch and that the direction shown is accurate to plus or minus 20 seconds. (This is consistent with a "Class B" survey.)

Times reports that there are no visible indications of any easements or other encumbrances to the property except the apparent encroachment of a fence line, which is shown. The fence is not very old, according to Times, and it appears that Beckmann has fenced in more than he has "a right to." Times recommends that the boundary with Beckmann be settled before any purchase is made. The title insurance company reports that they will insure title to that portion of the property that is not in conflict (332 feet wide). Times also agrees to add the easement found during the title search to the plat.

The surveyor cannot, and should not, pass judgment on whether or not Beckmann has "a right" to the strip of land between the fence line and the deed line. Beckmann's deed may not reflect a written claim to the 18-foot strip, but there may have been a transfer of rights based upon acquisitive prescription.

Surveyors do not establish boundaries. Times has performed all the work that he is qualified to perform. He has uncovered and documented evidence of a conflict between the lines of possession and the written deed, thereby providing you with all of the pertinent facts about the parcel that you need in order to make a decision. Times cannot warrant that Beckmann's possession can be ended. The boundary is in doubt, and only an agreement between owners or a court decision can settle it.

You may purchase the property and attempt to settle the question about the boundary. You may require the owner to settle the question about the boundary before any purchase is made. You may decline to purchase the property. These options should be considered in consultation with *your* attorney, not the title insurance company's attorney. Times may appear on your behalf, testify to the facts discovered by him, offer an opinion about the boundary, and many other things, but he cannot establish the boundary contrary to Beckmann's possession or consent. The survey plat can only serve as documentation of the extent of the conflict.

10.3. EASEMENT SURPRISE

Let us assume that you are a real estate investor contemplating the purchase of a large building in an industrial area. The present owner purchased the property in 1965, and a copy of the survey performed at that time is shown in Sketch 47. The owner states that he has not made any improvements to the property since the purchase.

Referring back to the checklist, you determine the following:

1. The parcel is a part of a platted subdivision.
2. The note that all corners are 90 degrees indicates that the plat is a tangent plane projection.
3. The survey is dated and appears to conform to the standards of practice during the 1960s.
4. The survey was for the purpose of assisting in the documentation of a sale.
5. a. North is shown, but no bearings are shown.
   b. The description shown is based upon the subdivision plat and is sufficient to identify the plat.
   c. The date of the plat indicates that it is over 20 years old.
   d. The name and address of the surveyor are shown.
   e. The signature and seal of the surveyor are shown.
   f. The streets bounding the square are shown. This is a common and excepted means of defining the boundaries of a lot in a platted subdivision in many communities.
   g. The dimensions of all sides are shown. The fact that the tenths and hundredths of a foot are not shown may be the result of an attempt at brevity, as far as the lot boundary lines are concerned. The lack of fractions in building offset lines and dimensions is more likely to be the result of imprecise work.
   h. The angles at the corners are reported to be 90 degrees.
   i. The name of the client is shown.
   j. The certification is present and warrants that the drawing faithfully reflects what J. Regular discovered during his survey.
   k. The certification contains the limiting words "According to the Recorded Subdivision Plan."
   l. The area is not shown.
   m. The scale of the plat is not shown.
6. The deed only refers to the parcel by naming the lot, square, subdivision, county, and state. The lack of a metes and bounds supple-

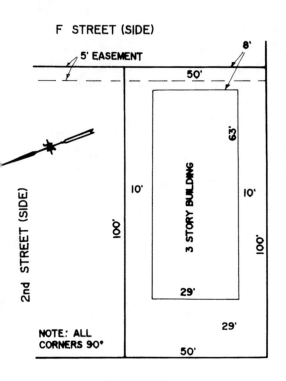

PLAT SHOWING
LOT I, SQUARE 36
OLD TIMES SUBDIVISION
SOME COUNTY
ANY STATE

F STREET (SIDE)

5' EASEMENT

8'

50'

63'

2nd STREET (SIDE)

10'

3 STORY BUILDING

10'

100'

100'

29'

3rd STREET

NOTE: ALL
CORNERS 90°

29'

50'

E STREET

Joe Regular

JOE REGULAR, P.L.S.   (SEAL)
2110 FALLS ROAD
ANYVILLE,
ANY STATE

TO: ATTORNEY JANE SMITH

MAY 10, 1965

I CERTIFY THIS PLAT TO BE A
TRUE AND ACCURATE REPRESENTATION
OF A SURVEY MADE BY ME ACCORDING
TO THE RECORDED SUBDIVISION PLAN.

SKETCH 47

mentary description removes any possible check for blunders in listing the identification of the lot.

7. The plat shows a 5-foot easement across the east side of the parcel.

8. There are no apparent encroachments.

9. The implied precision of the plat requires clarification.

10. The limiting phrase by Regular is notice that the *only* record that

Regular relied on for data concerning lot 1 was the recorded subdivision plan and the discoveries, measurements, and observations made by him during the survey.

11. When he was contacted to clarify the questions raised by the checklist, Regular confirmed that he did not perform any search of the deed record for anything other than information on the locations of the exterior boundaries of the lot. The subdivision plat indicated the easement line shown without reference to the party enjoying the rights of that easement. Regular reported all distances and angles to be consistent with the accuracy requirements of a "Class B" survey. The building is constructed of a rough, shingled exterior and is in disrepair. Regular recommended that you obtain title insurance, which you decline to do (it costs too much).

Unknown to all in this scenario was that a 10-foot easement had been acquired by a major gas service company along the entire east boundary of the lot. The easement was acquired only two years after the subdivision was created, one year before the construction of the building. The hidden 2-foot encroachment is a ticking time bomb that will hinder the marketability of this property as soon as it is discovered.

10.4. EXCESSIVE PROBLEMS

John Mason, a local developer and home builder, decided to build a "spec" house on lot 1, block C of East Kingston Heights subdivision, local county, our state. Mason is the owner and developer of both Kingston Heights subdivision and East Kingston Heights subdivision. Work progressed rapidly in both subdivisions. As part of the construction loan to Mason, the lender, Easy Money Inc., required that an ALTA/ACSM land title survey be performed at the end of construction. The work was completed on the house in July of 1986, and Sketch 48 shows the survey plat requested and used by Easy Money Inc., along with a complete title examination by Nit Pic Title Inc.

An application of the evaluation checklist to this plat reveals the following:

1. The land record system governing this parcel is the platted subdivision system.
2. The mapping system is probably some form of a conformal projection. It may be a tangent plane or state plane projection, but, for a

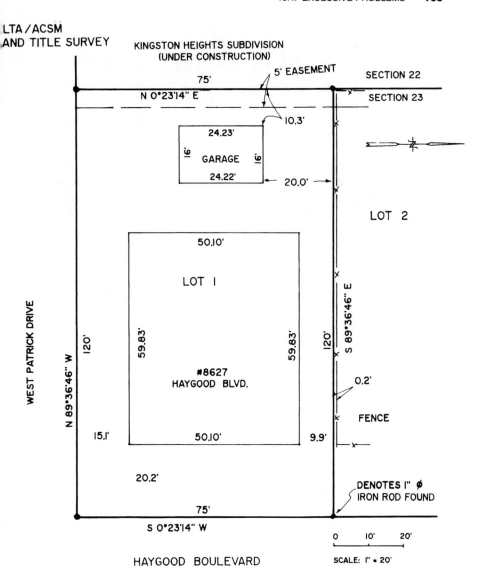

LTA / ACSM
AND TITLE SURVEY

KINGSTON HEIGHTS SUBDIVISION
(UNDER CONSTRUCTION)

SKETCH 48

THIS PLAT SHOWING LOT I, BLOCK C, OF EAST KINGSTON HEIGHTS
SUBDIVISION AS RECORDED IN BOOK 324 FOLIO 15, LOCAL COUNTY CLERK OF
COURT AND ALL BEING A PORTION OF SECTION 23, T8S R22E, BIG TOWN,
LOCAL COUNTY, OUR STATE HAS BEEN MADE BY ME AT THE REQUEST OF J. DOE
AND WAS PERFORMED TO THE ALTA / ACSM CLASS A STANDARDS.

H. T. SHOT, P.L.S.
2715 N. 3RD ST.
BIG TOWN, OUR STATE

JULY 22, 1986

SEAL

parcel this small, it is not very important to distinguish which one is in use.

3. The survey is modern, both chronologically and procedurally.

4. This survey was done for the purpose of fulfilling the boundary survey requirements of an ALTA/ACSM land title survey.

5. a. The north arrow is shown without a direct reference to a bearing base.

   b. The legal description is shown on the plat as "lot 1, block C, East Kingston Heights subdivision, as recorded in book 324 folio 15, local county, our state". This description is consistent with the land record system in force.

   c. The date of the survey is shown.

   d. The name and address of the surveyor are shown.

   e. The signature and seal of the surveyor are shown.

   f. The adjoining properties are shown as street rights-of-way, lot number, or section number.

   g. The dimensions of all sides are shown.

   h. The bearings of all sides are shown.

   i. The name of the client and the purpose of the survey are stated.

   j. The certification is not worded in exact accordance with the requirements of an ALTA/ACSM land title survey.

   k. The reworded certification contains the implication of limiting the records search by H. T. Shot to the recorded subdivision plat.

   l. The area is not shown.

   m. The scale is noted to be "1" = 20'" and is accompanied by a bar scale.

6. The subdivision is new, and a prior deed for this particular lot did not exist. The survey plat of lot 1 block C is consistent with the subdivision plat.

7. There is an easement across the entire western end of the lot. The title examination did not reveal any other easements or other noteworthy items of record.

8. There are no encroachments. The fence along the northern boundary of the lot is not considered by Nit Pic Title Inc. to be an encumbrance.

9. The survey standard of "Class A," as defined by the "Minimum Standard Detail Requirements for ALTA/ACSM Land Title Surveys," is of the highest order of accuracy for boundary surveys.

10. The certification should be changed to meet the ALTA/ACSM standard certification.
11. This step was not taken by Easy Money Inc.

Mrs. Gracie A. McCoy purchased this house in August of 1986. The mortgage company that financed this purchase, Cut Rate Loans Inc., used the July 22, 1986 survey by H. T. Shot, P.L.S. to document this sale. Nit Pic Title Inc. has the local reputation of being extremely meticulous, Easy Money Inc. is known to be a very cautious lender, and H. T. Shot's reputation is excellent. Cut Rate Loans did not require a new title examination, nor was an updated survey requested.

Eventually, Kingston Heights subdivision was also completed. The house and lot directly west of 8627 Haygood Blvd. was purchased by A. M. Schorr in January of 1987. On February 2, 1987, McCoy and Schorr agreed mutually to fund a fence along the line dividing their properties. The survey marks on Schorr's eastern boundary were visible, so the fencing company used these marks to build a wood fence along Schorr's eastern line.

McCoy became dissatisfied with the size of her garage, razed it, and rebuilt the garage to the dimensions of approximately 24 feet × 24 feet. Mrs. McCoy became dissatisfied with the entire house and sold it to R. R. Robert. This sale was also financed by Cut Rate Loans. The appraiser for Cut Rate Loans inspected the property in May of 1987 and noticed that the new garage was over 7 feet east of the west fence line. No other construction had taken place. No survey was required by the buyer or the lender for this new sale.

In August of 1988, R. R. Robert agreed to sell this same house (at a great profit) to Schorr's widowed mother-in-law, B. Axer. The loan for this sale was handled by Easy Money Inc., which immediately ordered a resurvey of this lot, in spite of the existence of a recent ALTA/ACSM land title survey on this property, dated July 22, 1986. This survey is shown in Sketch 49.

The reader may evaluate this survey plat, as well as contemplate the many solutions to the problem presented. The savings realized by not requiring a boundary survey for each exchange of title cannot compare with the losses that will result from the complicated lawsuit that Robert is about to bring against all involved. The rule of thumb in real property transfers is, "The land survey is too expensive when it is more than the cost of possible losses or the value of the land."

10.5. METES MEETS BOUNDS

Abraham Savings and Loan (AS&L), located in Capital City, east state is in the process of providing a mortgage loan to Mr. William Cheatham. Mr.

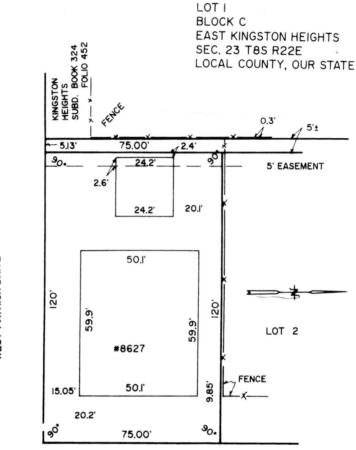

LOT 1
BLOCK C
EAST KINGSTON HEIGHTS
SEC. 23 T8S R22E
LOCAL COUNTY, OUR STATE

HAYGOOD BLVD.

I CERTIFY THIS PLAT TO BE A TRUE AND ACCURATE REPRESENTATION OF A
SURVEY MADE BY ME AT THE REQUEST OF WILLIAM BEES, ATTY.

AUG. 15, 1988

NOTE: THIS SURVEY IS BASED UPON THE RECORDED SUBDIVISION
PLAT AND INFORMATION PROVIDED BY THE CLIENT. NO SEARCH
OF THE TITLE RECORD WAS MADE BY THIS OFFICE.

BENJAMIN DERE, P.L.S. #101
2719 N. 3RD ST.
BIG TOWN, OUR STATE

SEAL

SKETCH 49

Cheatham provided a copy of a survey plat, showing the property to be mortgaged. Sketch 50 shows that survey plat. You are a loan officer with AS&L and have been assigned this loan.

Using the checklist from Chapter 9, you discover the following:

1. The land record system is a metes and bounds system. Two separate parcel histories (Johnston estate and Mills tract) must be examined for title certification in order to insure prime lien on the property. A copy of the survey plat is sent to the title insurance company, along with a request for insurance.

2. This is clearly a tangent plane projection survey.

3. The precision of the work seems to be modern.

4. The purpose of the survey appears to be documentation that the house under construction is within the property lines. The fact that a house is shown as "under construction" may be of serious concern if this is to be a first mortgage.

5. a. The north arrow is shown, and the bearing base is defined as the "call" direction for the Fairgrounds Highway (assumed north).

    b. The legal description is "the William Cheatham Property, Dade County, east state," along with the bounds shown.

    c. The date is shown.

    d. The name of the surveyor is shown, but no address is given.

    e. The signature and seal of the surveyor are shown.

    f. All adjoiners are shown.

    g. All dimensions are shown.

    h. All directions are shown.

    i. The name of the client is shown.

    j. The certification is very general and gives no hint about the intended purpose of the survey.

    k. There are no limiting words or phrases.

    l. The area is not shown.

    m. The scale is not shown.

6. The deed presented to you by Cheatham describes the property as being bounded by Riverside Drive, Fairgrounds Highway, the property of B. B. Mills, and the property of P. Petty, and more particularly described as:

> . . . Commencing at the corner of Fairgrounds Highway and Riverside Drive and True Point-of-Beginning; thence, South 75 degrees East along Fairgrounds Highway, a distance of 32 feet no inches; thence,

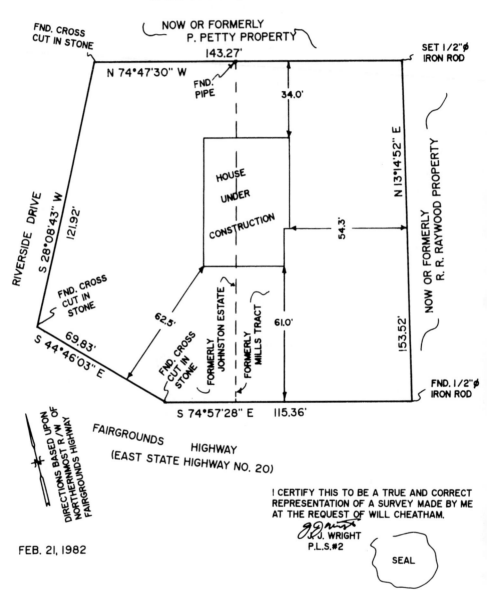

PLAT SHOWING THE
WILLIAM CHEATHAM PROPERTY
DADE COUNTY, EAST STATE

NOW OR FORMERLY
P. PETTY PROPERTY

FND. CROSS
CUT IN STONE

143.27'

N 74°47'30" W

SET 1/2"∅
IRON ROD

FND.
PIPE

34.0'

RIVERSIDE DRIVE

S 28°08'43" W   121.92'

N 13°14'52" E

NOW OR FORMERLY
R. R. RAYWOOD PROPERTY

HOUSE

UNDER

CONSTRUCTION

54.3'

FND. CROSS
CUT IN
STONE

62.5'

S 44°46'03" E   69.83'

FND. CROSS
CUT IN
STONE

FORMERLY
JOHNSTON ESTATE

FORMERLY
MILLS TRACT

61.0'

153.52'

FND. 1/2"∅
IRON ROD

S 74°57'28" E   115.36'

FAIRGROUNDS   HIGHWAY
(EAST STATE HIGHWAY NO. 20)

DIRECTIONS BASED UPON R/W OF
NORTHERNMOST HIGHWAY
FAIRGROUNDS HIGHWAY

FEB. 21, 1982

I CERTIFY THIS TO BE A TRUE AND CORRECT
REPRESENTATION OF A SURVEY MADE BY ME
AT THE REQUEST OF WILL CHEATHAM.

J. WRIGHT
P.L.S.#2

SEAL

SKETCH 50

North 13 degrees 15 minutes East, a distance of 153 feet no inches; thence, North 75 degrees West for 65 feet no inches; thence, fronting on Riverside Drive 122 feet no inches and a second fronting of 70 feet no inches to the point-of-beginning. . . .

The deed, though poorly written, clearly describes only a portion of the property depicted on the survey plat. No reference is made to the inclusion of the Mills tract.

7. No easements are shown.

8. No encroachments are shown onto or from the Cheatham property.

9. Accuracy standards are not stated.

10. Discrepancies between deed and the survey plat are profound and must be explained.

11. A telephone call to the surveyor (the east state board of registration provided an address and a telephone number) reveals that the survey plat was prepared to assist in the resubdivision of the Mills tract and the Johnston estate into one parcel. The local county ordinances require the introduction of a plat of resubdivision, along with proof of ownership by the parties requesting the resubdivision. The county, upon approval, records a signed copy of the survey plat as the official plat of resubdivision.

The title insurance company reported that Cheatham's title to the Johnston estate was free and clear. The Mills tract was purchased by Cheatham Enterprises, Inc. and carries a mortgage in favor of First Bucks International for $125,000. There is no record of a resubdivision combining the two parcels.

10.6. THE SQUARE THAT WASN'T THERE

Two years ago, you were requested by Cut Rate Loans Inc., a mortgage company, to examine the title on a parcel of land identified as "lots 1 through 22, square 16 Grandsire Farms, section 8, T4N R21W, this county, some state." The purpose of the title examination was for an act of sale, 70 percent financed by Cut Rate Loans, from the estate of O. Fogey to B. C. Jingwaski. Fogey acquired the property in 1908 from Night Investments, Inc., which had purchased all of section 8, T4N R21W from the U.S. government in 1899. Sketch 51 shows the subdivision plat of Grandsire Farms that was attached to the 1908 conveyance.

The subdivision of Grandsire Farms had only been partially developed at the time. Some streets had been improved, some squares had been resubdi-

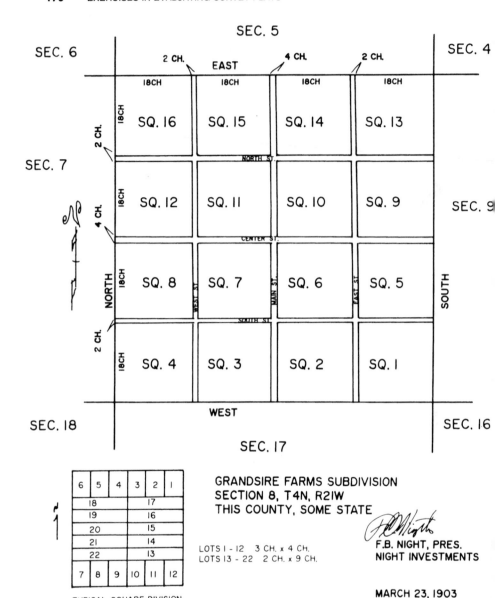

SKETCH 51

vided, and some areas were untouched woodlands. Jingwaski acknowledged that he was aware that square 16 and the surrounding street rights-of-way were heavily wooded and undeveloped. This area had become a rapidly developing suburban area of Cin City, and the sale price for square 16 was $400,000.

The examination of the conveyance record revealed only two exchanges involving square 16: The first was the sale by the U.S. government to Night Investments of all of section 8, and the second was the sale by Night Investments to Fogey of lots 1 through 22, square 16 "being all of Square 16." The record was as clean as you had ever seen.

An evaluation of the subdivision plat, using the checklist from Chapter 9, provides the following:

1. The land record system is a recorded subdivision plat.
2. The presence of cardinal directions (north, south, east, and west) and the lack of any indication that angles were measured indicate that this is a projectionless map.
3. The plat is dated March 23, 1903, but, because all of the distances are to the nearest chain and all of the directions are cardinal, you suspect that the last work done on this parcel was by a U.S. government surveyor prior to 1899.
4. The purpose of the plat (there was great doubt that a survey was ever conducted) is to create Grandsire Farms subdivision.
5. a. The north arrow is without reference to a bearing base.
   b. The legal description is indicated by the numbers in designated lots and squares. The legal description of Grandsire Farms is on the face of the plat.
   c. The plat is dated 1903.
   d. The name and address of the "surveyor" are shown.
   e. There is a signature but no seal.
   f. Adjoining properties are shown.
   g. All dimensions are shown. The exterior dimensions are identical to a regular USPLS section.
   h. Bearings are shown on the section boundaries only.
   i. The name of the client is shown, and the purpose of the plat is obvious.
   j. There is no certification.
   k. There are no limiting words or phrases.
   l. No areas are shown.
   m. There is no scale shown on the plat.
6. The deed only refers to lot and square numbers. There is no supplemental description.
7. Street rights-of-way are the only features shown on the plat.
8. No improvements are shown. You are familiar with the area and know that it is free of any fences, buildings, or cultivation.

9. You feel that $400,000 worth of land in a rapidly developing area adjacent to Cin City calls for a "Class A" survey.[1]
10. The 1903 subdivision plat does not fulfill very many of the requirements of an ALTA/ACSM land title survey.
11. The "surveyor" has been dead for 32 years.

You advised Cut Rate Loans that you would require that a new survey be performed on the property being sold. Your surveyor quoted a cost of $5,450 to perform an ALTA/ACSM land title survey. Cut Rate Loans and Jingwaski both knew that the land was heavily wooded and completely vacant. Both decided that there was no possibility of any adverse possession and declined to commission a survey. (Who ever heard of paying $5,500 for some guy to stick pipes in the ground in the middle of nowhere?)

Last week, Bigbucks Development, Inc. signed a purchase agreement with Jingwaski for all of square 16 Grandsire Farms for the sum of $750,000. Bigbucks Development contracted with a surveying and engineering firm to survey square 16 and to design a single-family homesite development. Bigbucks Development directed the surveyor to send the survey of square 16 to you to be used in the title examination for their purchase from Jingwaski. Today, the surveyor, C. W. Brushcut, presented you with sketches 52 and 53. Sketch 52 shows the U.S. township plat for township 4 north, range 21 west. Sketch 53 shows his survey plat of square 16.

"Call me when you have a chance," said Brushcut as he left.

[1] See Table 1 in Appendix A, "Minimum Standard Detail Requirements for ALTA/ACSM Land Title Surveys."

T4N   R2IW

Nth. MERIDIAN
SAME STATE

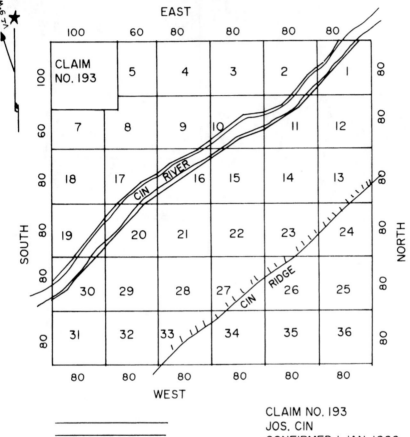

CLAIM NO. 193
JOS. CIN
CONFIRMED I JAN. 1888

APPROVED:

SURVEYOR GENERAL

SKETCH 52

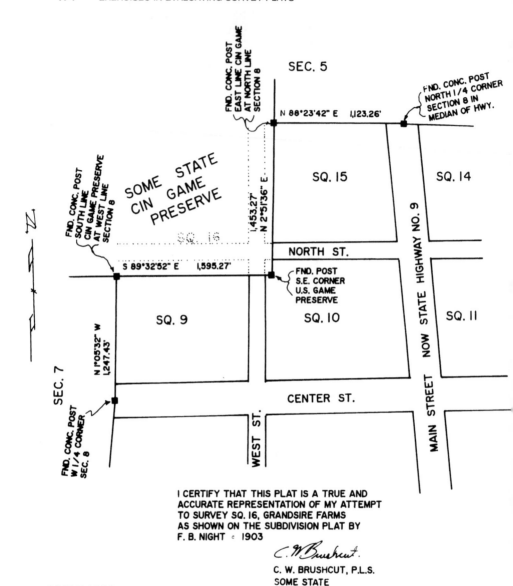

FND. CONC. POST
EAST LINE CIN GAME
AT NORTH LINE
SECTION 8

SEC. 5

FND. CONC. POST
NORTH 1/4 CORNER
SECTION 8 IN
MEDIAN OF HWY.

N 88°23'42" E   1,123.26'

SOME STATE
CIN GAME
PRESERVE

SQ. 16

SQ. 15

SQ. 14

FND. CONC. POST
SOUTH LINE
CIN GAME PRESERVE
AT WEST LINE
SECTION 8

1,453.27'   N 2°51'36" E

NORTH ST.

S 89°32'52" E   1,595.27'

FND. POST
S.E. CORNER
U.S. GAME
PRESERVE

SEC. 7

N 1°05'32" W
1,247.43'

SQ. 9

SQ. 10

SQ. 11

CENTER ST.

WEST ST.

MAIN STREET   NOW STATE HIGHWAY NO. 9

FND. CONC. POST
W 1/4 CORNER
SEC. 8

N

I CERTIFY THAT THIS PLAT IS A TRUE AND
ACCURATE REPRESENTATION OF MY ATTEMPT
TO SURVEY SQ. 16, GRANDSIRE FARMS
AS SHOWN ON THE SUBDIVISION PLAT BY
F. B. NIGHT ◦ 1903

C. W. Brushcut.

C. W. BRUSHCUT, P.L.S.
SOME STATE

JUNE 7, 1990

SKETCH 53

CHAPTER 11

WRITING LEGAL DESCRIPTIONS

The definitions of "legal description" and "land description" are sometimes thought to be interchangeable. However, there is a subtle difference between the two, and neither definition is exactly what the modern use of the phrase has come to mean. A legal description is one that is sufficient to locate the property without oral testimony.[1] A land description is one that defines the exact location of a parcel, stated in terms that are consistent with the land record system used to create the parcel, such as lot, square, subdivision, metes and bounds or aliquot part, section, township, range, and principal meridian. All land descriptions are legal descriptions.

The use of the phrase "legal description" has evolved to mean the body of words and phrases used within a real property deed to depict a specific parcel of land in a way that identifies that parcel, not only in terms of the land record system used to create the parcel, but also in terms of delineating the size, shape, and location of the parcel so specifically that recovery of the boundaries of that parcel is possible using only the written description. This particular "legal" or "land" description takes the same form in every land record system and, because we will only concern ourselves with descriptions within deeds, will be called the "deed description" hereafter.

This chapter will be devoted to developing the art of writing deed descriptions that are clear, complete, concise, and consistent (with the land record system in use and the intentions of the parties creating the title). When

---

[1] American Congress on Surveying and Mapping and the American Society of Civil Engineers, DEFINITIONS OF SURVEYING AND RELATED TERMS (rev. 1978)

called upon to determine intent, courts will consider all of the evidence presented in the deed description, as well as other evidence.[2] A well-written deed description will reflect the intent of the parties and will assist some future court in its deliberations. There will only be one recommended form of deed description covered in this chapter, and no attempt will be made to discuss the portions of the deed that cover the specific rights conveyed, conditions of the sale, and so forth. Many other forms of deed descriptions in use across the United States are very regionally oriented and are being gradually replaced by this more universal and flexible form. The method proposed here will enable persons with little or no trigonometric or geometric background to write excellent deed descriptions.

The first step in writing a deed description is to evaluate the intentions of the parties to the transaction. In most cases, this intent is to transfer certain rights to an existing parcel. In other cases, a new parcel is being created. Each of these situations has its own particularities that may alter slightly the composition of the deed description.

11.1. CREATING NEW PARCELS

Divisions of a real property into two or more new parcels, combining existing parcels into one, or rearranging the boundaries of existing parcels are all acts of resubdivision. A resubdivision is most easily accomplished by a complete land survey and the development of a subdivision plat by a surveyor. Although other land record systems allow for resubdivision of land, the platted subdivision accomplishes this so clearly and effectively that most local city and county governments now require platted subdivisions for all divisions of real property within their jurisdictions.

A new survey should be performed to collect the information about the property needed in order to decide upon division. Old survey plats will not reflect accurately present conditions or discover changes in boundaries. Current information is vitally important. Without it, the parties to the transaction may be misled. Divisions of real property based upon ancient data risk severe conflicts when, at a later date, the resulting parcels are possessed and it is discovered that the parent tract was not the size and shape it was believed to be. The creation of a platted subdivision affords the owners an opportunity to discover and to correct any dimensional discrepancy that may have existed in the old deed information. Once in possession of complete and accurate information on the actual present configuration of the parent

---

[2] J. S. Grimes, *Clark on Surveying and Boundaries,* 4th ed. (Bobbs-Merrill Co., Inc.: Indianapolis & New York, 1976).

tract, the parties involved in the resubdivision can form their intent and take action without excessive risk of conflict among the assigns.

11.2. EXISTING PARCELS

Most often, a real property transaction involves an existing parcel that is already described in a deed. The current deed may, or may not, accurately describe the property. Changes in boundary lines may have occurred, or updated information about the boundaries may render portions of the old deed information obsolete. If a deed is to report accurately all of the rights transferred, then it is important that these rights be known. The ALTA/ACSM land title survey procedure[3] is designed to discover all of these rights and to detect possible changes.

11.3. GENERAL OUTLINE

All deed descriptions take the form of a five-part essay consisting of (1) caption, (2) narrative, (3) deletions or additions, (4) summation, and (5) references. Part (2) may be absent if information, beyond the caption, is not known. Parts (3) and (5) may be absent, and the order of (3), (4), and (5) is sometimes rearranged. Nevertheless, deed descriptions that fulfill all of the five parts have proven to be the least likely to be misinterpreted at a later date.

11.4. THE CAPTION

The caption usually takes the form of identification that is most basic to the land record system appropriate to the parcel. In USPLS and platted subdivisions, the caption is always distinct from the rest of the description. In the metes and bounds system, the caption and the narrative may merge if caution is not exercised. To avoid this possibility, the phrase ''and more particularly described as follows:'' or words to that effect, is sometimes added at the end of the caption to alert the reader that the narrative section of the deed description is about to begin. Here are some examples of possible captions.

---

[3] See Appendix A.

That parcel of land designated as Lot 1, Square 12, Timber Estates, Waterproof, Louisiana, more particularly described as follows:. . . (platted subdivision)

That parcel of land known as the "Brown Estates" located in Winston County, Kentucky bounded on the north by J. R. Westly, on the east by Philip Jones, on the south by D. Davis and on the west by Kentucky State Highway Number 5, more particularly described as follows:. . . . (metes and bounds)

That parcel of land designated as the NE¼ of Section 2, T5S R4E, Fourth Principal Meridian, Hayseed County, Illinois, described as:. . . . (USPLS)

The common factor in each of these examples is that the caption alone is sufficient to identify the parcel without oral testimony, based upon the prevailing land record system. The land record system in the examples is noted within parentheses following the caption. The identification in the caption is not normally negated by a contradiction occurring in the narrative portion of the deed description, unless it can be shown that the caption contains an error. If, for example, in a platted subdivision, the caption identifies a parcel by lot, square, and subdivision, according to a recorded subdivision plat, and the narrative portion reports a contradictory distance to a street corner, the caption (and the recorded plat that is an integral part of the caption by reference) will usually be assumed to be correct, unless additional evidence to the contrary has been found.

11.5. THE NARRATIVE

The narrative portion of the deed description always takes the form of that portion of a complete metes and bounds description, beginning with a commencement point, through a point-of-beginning, around the parcel, and returning to the point-of-beginning. The distinctions of calls for specific boundaries, bearings, distances, landmarks (monuments), and other distinctive factors discussed in Chapter 7 dealing with metes and bounds descriptions apply in a deed description. The narrative portion of a deed description is the same for all land record systems and is dependent upon a survey having been performed on the parcel being described. Divisions of property that take place without a survey, such as the aliquot division of a section, should not have a narrative portion in the deed description. Not all factors of a metes and bounds description are found in the narrative portion of a deed description, and, on the off chance that the reader did not read that section or does not remember it, the key factors will be repeated here.

The metes and bounds narrative must contain the following:

1. A commencing point that is well known, easily found, durable, recoverable, recognizable, and preferably public in origin.
2. A point-of-beginning that is a distinct part of the parcel being described.
3. A report of the physical objects (monuments) that mark the location of the ends or are found along each line.
4. A report of the owners or parcel identifiers of the contiguous properties along each line.
5. A direction for each line. This can be in the form of a bearing, an angle, or, in the case of curved lines or meandering lines, a description of the configuration of the line.
6. A length of each line segment, usually in the form of a distance between corners.

There are several recommended key words or phrases that should be used in the narrative of a deed description that are intended to assist the reader in identifying which of the preceding items is about to be reported or to clarify some other aspect of the description. Custom in a particular area may mean that a different combination of key words or phrases will be used to express the intent of the parties. In these areas, the introduction of the recommended system should be made gradually or with slight modifications for the sake of continuity.

How corner monumentation is identified is also significant in illustrating the intentions of the parties. Each region has developed certain preferred objects used to monument corners. Terrain, agriculture, local industries, geology, and other factors unique to an area all influence the kinds of monumentation of real property corners practiced by the landowners.

The format recommended here is easily adapted to a wide variety of regional customs and practices involving identification, monumentation, and description of real property. Once this format becomes a habit, reading or writing the narrative portion of a deed description will become the simplest part of working with real property identification and boundaries.

11.6. KEY PHRASES

11.6.1. "Commencing at"

"Commencing at" normally are the first two words of the narrative portion of a deed description. The reader is alerted to the fact that items following this phrase will be a physical object and/or a distinct theoretical land point that, in the opinion of the writer, fulfills most, if not all, of the requirements enumerated for commencement points.

The commencement point must be just that—a land point. It would not be proper to state, for instance, "Commencing at the northernmost right-of-way line for U.S. Highway 66." The reader is not told where along the northernmost right-of-way line the land point is located. It would be correct to state "Commencing at the intersection of the northernmost right-of-way line of U.S. Highway 66 and the easternmost right-of-way line of U.S. Highway 15." The second version clearly designates a land point that any reader of the deed could easily find. Physical objects that monument the commencement point should also be "called for" in the narrative, as well as any coordinates or other pertinent data.

11.6.2. "Point-of-Beginning"

The point-of-beginning brackets the narrative that traces or "traverses" along the boundaries of the parcel being described. This statement prepares the reader to begin sketching a mental picture of the parcel. The point-of-beginning is fundamental to the narrative portion of a deed description and, as with the commencement point, must be a true land point. Physical monuments, as well as theoretical land points, can serve as points-of-beginning. It is quite common for the monument, theoretical location, and state plane coordinates of a point-of-beginning all to be present in the narrative.

Although it is sometimes combined with the commencement point, the point-of-beginning cannot be dispensed with. "Commencing and point-of-beginning at . . ." is used when both are the same point, as in a corner lot. The second appearance of the words "point-of-beginning" in a narrative announces the termination of the traverse encompassing the parcel. When it is mentioned the second time, there is no further elaboration of any distinctive features of the point-of-beginning.

11.6.3. "thence," and ";"

A deed description is similar to a Faulkner sentence. From caption to references, the entire essay comprises one sentence. The word "thence" is used to announce that a new line segment is about to begin. The boundary line segment description always ends with a semicolon ";" giving the phrase that describes a single boundary line segment the appearance: "thence . . . ; . . . ." In this way, the very long narrative can be divided into "digestible" pieces that stand out to the reader. It is important to reserve these two items for this particular use. Applications of semicolons or the word "thence" for another purpose anywhere else in the narrative portion of a deed description will certainly cause confusion.

11.6.4. "in a ———ly direction"

The eight general directions of "norther*ly*," "northeaster*ly*," "easter*ly*," "southeaster*ly*," "souther*ly*," "southwester*ly*," "wester*ly*," and "northwester*ly*," are especially useful to persons who are not particularly familiar with reading and interpreting bearings. Directions with the suffix "ly" eliminate confusion over which form of a bearing is applicable for describing a particular line. Examine Sketch 54.

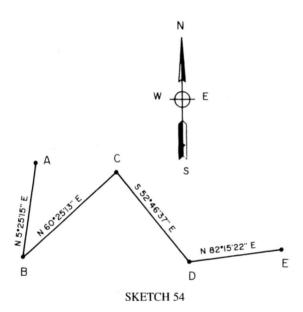

SKETCH 54

If the traverse being described includes the line segment from point "A" to point "B," an examination of the north arrow clearly indicates that this requires a move to the south. The phrase "in a *southerly* direction" prepares the reader to expect the bearing "*south* 5 degrees 25 minutes 15 seconds *west*" instead of "north 5 degrees 25 minutes 15 seconds east," which is printed in Sketch 54. Similarly, from point "B" to point "C" would call for the use of the phrase "in a *northeasterly* direction" to indicate that the bearing to use is "*north* 60 degrees 215 minutes 13 seconds *east*." The line segment from point "C" to point "B" would call for the phrase "in a *southwesterly* direction" to indicate that the bearing to use for that direction is "*south* 60 degrees 25 minutes 13 seconds *west*."

Notice that, if in describing the line from point "C" to point "B" the writer had decided to use "in a *southerly* direction" instead of "in a *southwesterly* direction," the choice of bearings would still be "*south 60 degrees 25 minutes 13 seconds west.*" The selection of "in a *westerly* direction" also results in "south 60 degrees 25 minutes 13 seconds *west.*" The use of these general directions allows a great deal of freedom of choice. The only limitation in choosing an "ly" direction occurs when the lines are close to the cardinal directions. Lines that are nearly north–south should use "northerly" or "southerly"; lines that are nearly east–west should use "easterly" or "westerly."

11.6.5. "-most"

The suffix "most" is used with the same eight general directions as the suffix "ly." The purpose of the general directions is to distinguish between items in a way that is obvious by an inspection of the property or its survey plat. As with the "ly" directions, there is wide leeway given in the choice of directions using the suffix "most." The majority of occasions that require these words can be satisfied with the cardinal directions. In Sketch 55, "northernmost," "northeasternmost," "easternmost," and "southeasternmost" all indicate the Jones and Smith side of U.S. Highway 77. Of the four choices, "northeasternmost" and "easternmost" are the most obvious, and equally preferred, selections in this example.

11.6.6. "along"

The key word "along" indicates that the boundary line being described is congruent with some other previously established line. Whether the previously established line is straight, curved, or meandering, the key word "along" indicates that the boundary line being described is exactly superimposed on the previous line at every point. For example, ". . . ; thence, in a southerly direction, *along the line between section 5 and section 6 . . . ;*" declares the described boundary to be the section line, and ". . . ; thence, in a northerly direction, *along the bank of the Snake River . . . ;*" leaves no doubt that this is a riparian boundary.

11.6.7. "a distance of"

The phrase "a distance of" is sometimes considered to be unnecessary. However, deed descriptions that refer to distances that are not lengths of boundary line segments can become very confusing if this key phrase is eliminated. For example, examine the excerpt ". . . ; thence, in a southerly

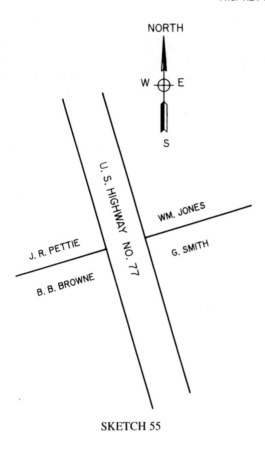

SKETCH 55

direction, along a curved line, having the radius center to the east and a radius of 200.00 feet, 350.00 feet. . . .'' or ''. . . ; thence, in a southerly direction, along a curved line having a radius of 200.00 feet and the radius center to the east, 350.00 feet . . .'' is not as clear as ''. . . ; thence, in a southerly direction, along a curved line, having the radius center to the east and a radius of 200.00 feet, *a distance of* 350.00 feet. . . .''

11.6.8. "to" and "to a point"

The key word ''to'' has special significance in demonstrating the intention of the parties in a transfer of titles. ''To'' announces that the boundary line segment is being terminated and that the corner description that *immediately* follows the word ''to'' is considered by the writer to control the length of the boundary line segment. The words ''to a point'' indicate that the limit is not set by a physical feature but by a theoretical location. Many other supplementary descriptions of the corner, or accessories to the corner, may follow the primary one. This distinction is best explained by example.

". . . 723.32 feet *to a point on the easternmost boundary of the Jones estate,* that same point being monumented by a 1-inch diameter iron pipe; . . ." clearly designates the "Jones estate" (an adjoining property) as the limit of that line segment. By using this phrase, the writer is stating that, if, for some reason, the 1-inch diameter iron pipe was found to be short of the Jones estate or the distance called for was insufficient to reach the Jones estate, the limit of the line segment would not be the pipe or the distance but would continue on to (or stop at) the easternmost boundary line of the Jones estate.

". . . 723.32 feet *to a point* monumented by a 1-inch diameter iron pipe; . . ." indicates that the intention of the parties was to set the length of the line at *a specific distance* (723.23 feet), and, if the pipe were later to be found to be short of, or in excess of, that distance, the distance would prevail (in the absence of possession or other convincing and contradictory evidence). In this instance, the pipe is just a marker and has not "achieved" the status of an artificial monument.

". . . 723.32 feet *to* a 1-inch diameter iron pipe; . . ." indicates, according to the intent of the writer, that the iron pipe is a artificial monument and that it, not the distance, is the controlling factor.

The exception to this "rule" occurs in cases in which real property boundaries are named in any part of the corner description. The mention of a real property boundary usually indicates the intention to limit the line at that real property boundary, even if calls for markers or other corner identification occurs first. In any case, the order is not exclusive and binding. As with any part of the deed description, other evidence of intent may be used to refute a contention based upon order of appearance of corner descriptions. In areas where the custom of writing deed descriptions is well established and does not follow the pattern that is recommended here, it may require some time to introduce this method.

11.6.9. "that same point"

The corner to a real property parcel is a theoretical location that may refer to several physical or theoretical things to define that location. Any corner may have several different items in the list of these things. A corner may be monumented by a surveyor's marker, may be part of a common boundary line, may have several accessories, may have known state plane coordinate values and other features that need to be described. The phrase "that same point" prevents the reader of a deed description from confusing the features of one corner with those of another. An example of the use of this phrase is ". . . to a stone borne located on the line between sections 30 and 31, *that same point* having the Virginia state plane coordinates (south zone) of

$y = 234,543.55$ feet, $x = 1,323,445.39$ feet, and point-of-beginning; thence, ...''

11.6.10. "point of curvature" and "point of tangency"

There are two types of intersections possible between lines when one or both of the lines are curved. A line—even another curved line—may simply intersect with a curved line, or it may meet that curved line in a very special way, known as a "tangent" intersection. Sketch 56 shows several combinations of intersections involving curved lines.

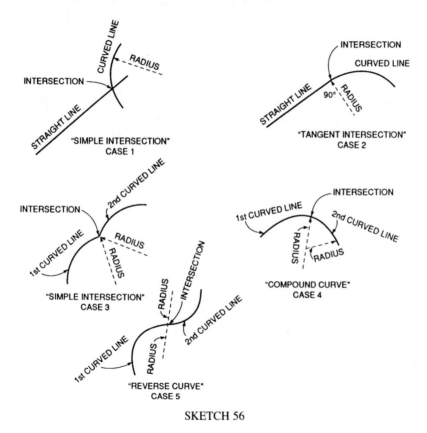

SKETCH 56

The situations labeled "tangent intersection," "compound curve," and "reverse curve" are all intersections where the radius center of the curve has a special relationship to the point of intersection. In the straight line tangent intersection (case 2), the radius center is located on a line from the intersection that is at right angles to the straight line. In the case of the compound curve and the reverse curve (cases 4 and 5), the radius centers are on a straight line drawn through the point of intersection. The occurrence of these sorts of intersections in deed descriptions requires the use of the phrase "point of curvature" when a straight line ends and a curved line begins at a point of tangency. The phrase "point of compound curvature" is used when a curved line of one radius intersects with a curved line of a different radius, but the same direction, at a point of tangency. The phrase "point of reversed curvature" is used when a curved line intersects with another line curving in the opposite direction at a point of tangency. Although all of the intersections discussed here occur at a point of tangency, the key phrase "point of tangency" is used in a deed description only when the description of a curved line ends at a point of tangency with a straight line.

11.6.11. "from whence bears"

The phrase "from whence bears" is used exclusively to designate the location of accessories to a corner. The particular wording of the segments that describe accessories is deliberately "old-fashioned" so that accessory descriptions do not become confused with boundary line segments. For example, ". . . to an iron post *from whence bears* a large oak at north 23 degrees east and 23 feet distant, . . ." makes the accessory a part of the deed description. In general, naming accessories in deed descriptions has not been routinely practiced.

11.6.12. "encompassing an area of"

The phrase "encompassing an area of" should be placed directly after the second occurrence of the words "point-of-beginning" to emphasize that the area expressed is that of the entire parcel and does not account for any changes in area caused by the deletions or additions noted later in the deed description. For example, ". . . to the point-of-beginning; encompassing an area of 234.43 acres . . ." leaves no doubt that the narrative is believed to describe a parcel containing that area. If the statement of area is removed from the narrative portion of the deed description, then confusion can occur between area before and area after accounting for the deletion or addition.

11.7. DELETIONS OR ADDITIONS

The deletions or additions portion of a deed description usually lists all of the easements, servitudes, rights-of-way, restrictions, or other modifications to the free, clear enjoyment of the parcel described. Sometimes, more commonly in rural areas, this portion of the description actually entirely eliminates portions of the described tract from the deed. This practice can cause confusion and is becoming less common as local governments enact stricter and more formal codes or ordinances controlling the resubdivision of real property. The reader is urged to resubdivide and plat any deletion or addition to a parcel that completely changes all of the real property rights of that deletion or addition.

This deletions or additions portion of a deed description is clearly set apart from the narrative portion by the use of, in bold or capital letters, the heading "LESS THAN AND EXCEPT:" "IN ADDITION:" or "SUBJECT TO:" These are not found in every deed description. Unfortunately, if this heading appears on a separate page from the narrative portion of the deed description, it may become separated from the rest of the deed documents. For this reason, it is recommended that the summation and references always follow the deletions or additions portion of the deed description. If this procedure is followed, the reader will always know that a complete deed description is present if a caption to the references is present.

Some examples of the use of deletions or additions in deed descriptions follow.

"Subject To:

A 5-foot easement adjoining and along the entire rear or northerly portion of the property for the purpose of installation and maintenance of a telephone cable. . . ."

"Subject To:

An easement of ingress and egress, 20-feet wide, adjoining and along the western boundary for the use of the owner of lot 3 and more particularly described as follows;. . . ."

"Less Than and Except:

That portion of the property known as "Devils Well" and more particularly described as follows;. . . ."

11.8. REFERENCES

The references portion of the deed description is used to "call for," thereby including in the deed, other documents, such as a survey plat, subdivision plat, maps, transactions, and the like. Items so included in the references will usually be considered as appearing in the deed description in full. A reference to a survey plat will normally have the effect of causing everything on the plat to become an integral and valid part of the deed. Sometimes the list of previous exchanges involving the parcel is also made in the references. Some examples of references follow.

> "... and all as more fully described on the plat of survey by T. T. Thomas, P.L.S., dated February 2, 1986."
>
> "... being the same property as acquired by the present owner from Sunshine Realty on December 4, 1944 and recorded in. ..."
>
> "... being the same property created by the subdivision plan of Oak Hills subdivision by T. Swift, L.S. and recorded in. ..."

The purpose of a deed description is to express clearly the intent of the parties conducting the exchange of title. Information that is relevant to that intent ought to appear in the deed description in one of the categories already mentioned. A properly written deed description will enable courts that are called on to settle a boundary dispute to eliminate deed discrepancies as an object of contention. A little extra time and effort in writing deed descriptions will benefit any assigns or heirs by reducing confusion and litigation.

CHAPTER 12

EXERCISES IN WRITING DEED DESCRIPTIONS

The following exercises in writing deed descriptions will present a sketch of a fictitious real property parcel, special conditions, and the author's recommended deed description. In reviewing the recommended deed description, you may wish to evaluate the survey plat using the checklist on page 123 and to note whether the description reflects the circumstances accurately.

12.1. CASE 1

Bill Wright, the owner of section 11, T22N R43W, Hog County, farm state, wishes to sell, to Mike Farmer, the "southeast 160 acres" of section 11. Sketch 57 shows the only drawing of the section in Wright's possession. Wright acquired section 11 from Crystal Shine on November 10, 1942.

Farmer confirms that he wants to buy the "SE$\frac{1}{4}$ of Section 11" because he "a square plot 2,640 feet on a side." Of course, the "SE$\frac{1}{4}$," the "southeast 160 acres," and "a square plot 2,640 feet on a side" are three different divisions of section 11.

The deed description of the SE$\frac{1}{4}$ of section 11 can be made without a survey and would be "the SE$\frac{1}{4}$ sec. 11, T22N, R43W, Nth PM, Hog County, farm state and being a portion of the same property acquired by Bill Wright, the present seller, from Crystal Shine on November 10, 1942." It would be quite improper to quarter section 11 based upon the "ideal section" template and to add a narrative describing the SE$\frac{1}{4}$ as beginning at the section corner and measuring 40 chains (2,640 feet) to a side. If anything is certain

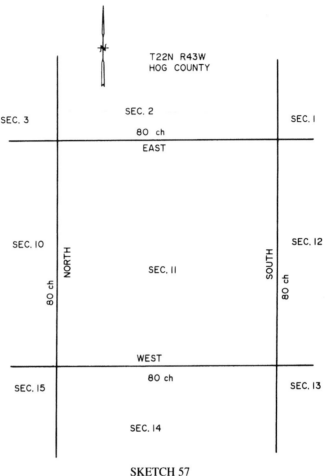

SKETCH 57

about section 11, it is that the section is *not* 80 chains north, east, south, and west. Farmer wisely orders a survey of the SE¼ of section 11. Sketch 58 is a rendering of that survey.

An abstract of the deed record (title) does not indicate any recorded easements, servitudes or rights-of-way. The surveyor, Ben Round, reports that South Road was built and has been maintained by the local farmer's co-op. The size of the right-of-way shown is based upon a written agreement on file at the co-op and is signed by Wright and others. Now the deed description can be written. Wright and Farmer both know exactly what is to be transferred. The description might read as follows:

That parcel or portion of land designated as the SE¼ of section 11, T22N, R43W, Nth PM, Hog County, farm state, and more particularly described as follows:

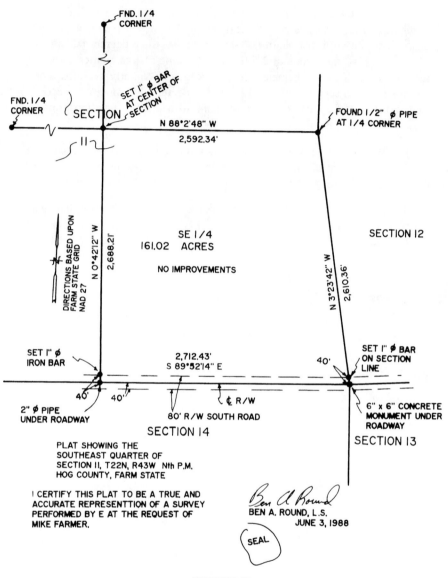

FND. 1/4
CORNER

FND. 1/4
CORNER

SET 1" ∅ BAR
AT CENTER OF
SECTION

SECTION

N 88°2'48" W
2,592.34'

FOUND 1/2" ∅ PIPE
AT 1/4 CORNER

DIRECTIONS BASED UPON
FARM STATE GRID
NAD 27

N 0°42'12" W
2,688.21'

SE 1/4
161.02   ACRES

NO IMPROVEMENTS

SECTION 12

N 3°23'42" W
2,610.36'

SET 1" ∅
IRON BAR

2,712.43'
S 89°52'14" E

40'

SET 1" ∅ BAR
ON SECTION
LINE

40'   40'

2" ∅ PIPE
UNDER ROADWAY

¢ R/W

80' R/W SOUTH ROAD

SECTION 14

6" x 6" CONCRETE
MONUMENT UNDER
ROADWAY

SECTION 13

PLAT SHOWING THE
SOUTHEAST QUARTER OF
SECTION 11, T22N, R43W Nth P.M.
HOG COUNTY, FARM STATE

I CERTIFY THIS PLAT TO BE A TRUE AND
ACCURATE REPRESENTTION OF A SURVEY
PERFORMED BY E AT THE REQUEST OF
MIKE FARMER.

Ben A Round
BEN A. ROUND, L.S.
JUNE 3, 1988

SEAL

SKETCH 58

Commencing and point-of-beginning at the southeast corner of section 11, that same point being monumented by a 6-inch square concrete monument; thence, in a westerly direction along the line between sections 11 and 14, that same line being the centerline of the South Road right-of-way, north 89 degrees 52 minutes 14 seconds west, a distance of 2,712.43 feet, to the southern quarter corner of section 11 and a 2-inch diameter iron pipe; thence, in a northerly direction along the line between the southeast quarter and the southwest quarter of section 11, north 0 degrees 42 minutes 12 seconds west, a distance of 2,688.21 feet, to the center of section 11 and a 1-inch diameter iron bar; thence, in an easterly direction along the line between the southeast quarter and the northeast quarter of section 11, south 88 degrees 02 minutes 48 seconds east, a distance of 2,592.34 feet, to the eastern quarter corner of section 11 and a ½-inch diameter pipe; thence, in a southerly direction along the line between sections 11 and 12, south 3 degrees 23 minutes 42 seconds east, a distance of 2,610.36 feet, to the point-of-beginning; encompassing an area of 161.02 acres:

SUBJECT TO: a right-of-way, 40 feet in width, adjacent to, and along, the entire southern line of the property herein described, for the purpose of the maintenance of a roadway:

All as more fully described on a plat of survey by Ben A. Round, land surveyor, dated June 3, 1988, and being a portion of the same property acquired by Bill Wright, the present owner, from Crystal Shine on November 10, 1942.

12.2. CASE 2

Iron Town, Bay County, Maryland, was created by a subdivision plan by S. Blake, county surveyor. The plan was dated May 3, 1920 and recorded in the Bay County conveyance office in book 5, page 2. Sketch 59 shows a

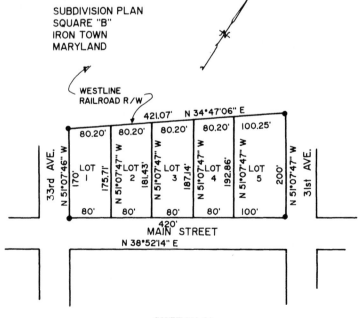

SKETCH 59

portion of that plan. Blake noted on the plan that "square corners were marked by a 2-inch diameter cast iron pipe." Write a deed description for lots 1 and 3, square "B."

The description of lot 1 might be as follows:

That parcel of land designated as lot 1, square B, Iron Town subdivision, Bay County, Maryland and all as more fully described as follows:

Commencing and point-of-beginning at the intersection of the northeastern-most right-of-way line of 33rd Avenue and the northwesternmost right-of-way line of Main Street, that same point being monumented by a 2-inch diameter cast iron pipe; thence, in a northerly direction along said northeast-ernmost right-of-way line of 33rd Avenue, north 51 minutes 07 minutes 46 seconds west, a distance of 170 feet, to the southeasternmost right-of-way line of the Westline Railroad, that same point being monumented by a 2-inch diameter cast iron pipe; thence, in an easterly direction along said southeast-ernmost right-of-way line of the Westline Railroad, north 34 degrees 47 min-utes 06 seconds east, a distance of 80.20 feet, to the line between lots 1 and 2; thence, in a southeasterly direction along said line between lots 1 and 2, south 51 degrees 07 minutes 47 seconds east, a distance of 175.71 feet, to the northwesternmost right-of-way line of Main Street; thence, in a southwesterly direction along said southeasternmost right-of-way line of Main Street, south 38 degrees 52 minutes 14 seconds west, a distance of 80 feet, to the point-of-beginning; all as more fully described on the plan of subdivision for Iron Town subdivision by S. Blake, county surveyor, dated May 3, 1920, recorded in book 5, page 2 of the Bay County conveyance office.

The description of Lot 3 might read as follows:

That parcel of portion of land designated as lot 3, square B, Iron Town sub-division, Bay County, Maryland, more particularly described as follows:

Commencing at the intersection of the easternmost right-of-way line of 33rd Avenue and the northernmost right-of-way line of Main Street, that same point being monumented by a 2-inch diameter cast iron pipe; thence, in an easterly direction along said northernmost right-of-way line of Main Street, north 38 degrees 52 minutes 14 seconds east, a distance of 160 feet, to the line between lots 2 and 3; thence, in a northerly direction along said line between lots 2 and 3, north 51 degrees 07 minutes 47 seconds west, a distance of 181.43 feet, to the southernmost right-of-way line of the Westline Rail-road; thence, in an easterly direction along said southernmost right-of-way line of the Westline Railroad, north 34 degrees 47 minutes 06 seconds east, a distance of 80.20 feet, to the line between lots 3 and 4; thence, in a southerly direction along said line between lots 3 and 4, south 51 degrees 07 minutes 47 seconds east, a distance of 187.14 feet, to the northernmost right-of-way line of Main Street; thence, in a westerly direction along said northernmost right-of-way line of Main Street, south 38 degrees 52 minutes 14 seconds west, a distance of 80 feet, to the point-of-beginning; all as more fully described on the plan of subdivision of Iron Town subdivision, by S. Blake, county surveyor, dated May 3, 1920, and recorded in the Bay County conveyance office in book 5, page 2.

12.3. CASE 3

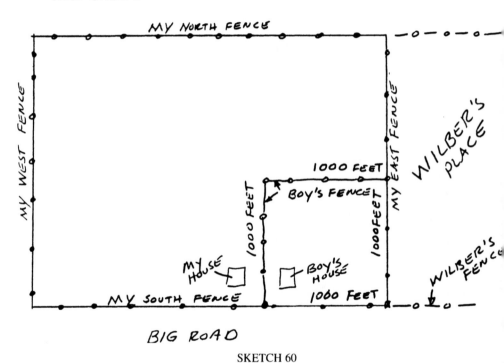

SKETCH 60

T. R. Grubb wants to give his son, R. T. Grubb, a portion of his farm property in Hardtack, Georgia. Sketch 60 is a drawing, made by the senior Grubb, of the property to be conveyed. The Grubbs refuse to have a survey made on the property, because "Dem survey folks don't know nuttin'. We gotta move dere markers every time dey come by." Write a description of the parcel conveyed.

Perhaps a description of the son's parcel would be as follows:

That portion of land located in Hardtack, Hardtack County, Georgia, bounded on the south by Big Road, on the east by the property of T. T. Wilber, and on the north and west by the property of T. R. Grubb, and more particularly described as follows:

Commencing and point-of-beginning at the intersection of the westernmost fence line enclosing the property of T. T. Wilber and the fence line along the northernmost side of Big Road; thence, in a westerly direction along said northernmost fence line of Big Road, a distance of 1,000 feet, more or less, to an intersecting fence line enclosing the property herein conveyed; thence, in a northerly direction, along said fence line enclosing the property herein conveyed, a distance of 1,000 feet, more or less, to an intersecting fence line enclosing the property herein conveyed; thence, in an easterly direction along said fence line enclosing the property herein conveyed, a distance of 1,000 feet, more or less, to the westernmost property line of T. T. Wilber and a fence line; thence, in a southerly direction along said westernmost property line of T. T. Wilber, a distance of 1,000 feet, to the point-of-beginning; and all as more fully described on a sketch by T. R. Grubb, hereto attached.

This last exercise demonstrates that the absence of a recent land title survey and a proper subdivision plan does not prevent the creation of deed description that almost appears adequate. Unfortunately, the distances are doubtful, the angles at the corners are unknown, the corners are poorly monumented, the lines are dependent upon possibly meandering fences, and the area is unknown. Pity the surveyor who, after 20 years and after the area has been cleared, is charged with the recovery of this particular parcel of land.

APPENDIX OF TABLES

TABLE 1   Select Units of Linear and Square Measure

1 U.S. survey foot	$= \frac{1200}{3937}$ meters
1 U.S. standard foot	= 0.3048 meters
1 line	= $\frac{1}{8}$ inch [a]
1 foot	= 12 inches
1 yard	= 3 feet
1 rod, pole, perch	= 16$\frac{1}{2}$ feet [b]
1 chain	= 66 feet
1 chain	= 100 links
1 chain	= 4 rods, poles, perches
1 mile	= 80 chains
1 mile	= 5,280 feet
1 nautical mile	= 6,080.27 feet [c]
1 rood	= $\frac{1}{4}$ acre
1 acre	= 43,560 square feet
1 acre	= 10 square chains
1 square mile	= 640 acres
1 compass, row	= 6 feet [a]
1 French foot	= 1.0657 feet [a]
1 toise	= $\frac{1}{30}$ arpent
1 toise	= 6.3944 feet [a]
1 arpent	= 191.994 feet (GLO definition)
1 arpent	= 191.83 feet
1 arpent square measure	= 0.8448 acres [a]
1 vara	= 33$\frac{1}{3}$ inches [a]
1 hand	= 4 inches
1 span	= 9 inches
1 cubit	= 18 inches
1 step	= 2$\frac{1}{2}$ feet [a]

TABLE 1 (Continued)

1 pace	= 5 feet[a]
1 fathom	= 6 feet
1 furlong	= 10 chains
1 league	= 3 miles[a]

[a] These units may vary greatly, depending on local custom.
[b] Feet listed in this table are U.S. survey feet, unless otherwise noted.
[c] This distance is a function of the spheroid in use and will vary. Some report this distance as 6.076.10 feet.

TABLE 2   Common Angular Units of Measure

1 circle	= 360 degrees
1 circle	= 400 grads
1 circle	= 32 points
1 circle	= 6,400 mils
1 degree	= 60 minutes
1 minute	= 60 seconds
1 degree of arc	= 17.45 feet in 1,000 feet[a]
1 minute of arc	= 0.29 feet in 1,000 feet[a]
1 second of arc	= 0.005 feet in 1,000 feet[a]
1 second of latitude at the equator	= 101 feet[a]
1 minute of latitude at the equator	= 6,046 feet[a]

[a] Approximate distance only.

TABLE 3 Accuracy Specifications and Positional Tolerances

Condition	A Urban	B Suburban	C Rural	D Mountain or Marshland	
Unadjusted closure (maximum allowable)	1:15,000	1:10,000	1:7,500	1:5,000	Traverse loop or between control monuments
Angular closure (maximum allowable)	10″ N	15″ N	25″ N	30″ N	N = number of angles in traverse
Accuracy of bearing in relation to source (maximum allowable)	± 15 sec.	± 20 sec.	± 30 sec.	± 40 sec.	1/sine angle = denominator in error of closure (approximately)
Linear distances accurate to: (maximum allowable)	± 0.05 ft. per 1,000 ft.	± 0.1 ft. per 1,000 ft.	± 0.15 ft. per 1,000 ft.	± 0.2 ft. per 1,000 ft.	Sine angle × 1,000 (approximately) where = accuracy of bearing
Positional error of any monument (maximum)	.03′ + AC/15,000	.05′ + AC/10,000	.07′ + AC/7,500	.1′ + AC/5,000	AC = length of any course
Calculation of area— accurate and carried to nearest ___ decimal place of acre	.0001	.0001	.001	.001	To 1 acre
	.001	.001	.01	.01	To 10 acres
	.01	.01	.1	.1	To 100 acres
	.1	.1	.2	.3	To 1,000 acres

TABLE 3 (Continued)

Condition	A Urban	B Suburban	C Rural	D Mountain or Marshland	
Elevations for boundaries controlled by tides, contours, rivers, etc., accurate to:	± .03 ft.	± .05 ft.	± 0.1 ft.	± 0.2 ft.	Based on accepted local datum
Location of improvements, structures, paving, etc. (tie measurements)	± 0.1 ft.	± 0.2 ft.	± 0.5 ft.	± 1 ft.	
Scale of maps sufficient to show detail but no less than	$1'' = 200'$	$1'' = 400'$	$1'' = 1,000'$	$1'' = 2,000'$	
Positional error in map plotting not to exceed: (applies to original map only)	5 ft. $1'' = 200'$	10 ft. $1'' = 400'$	25 ft. $1'' = 1,000'$	50 ft. $1'' = 2,000'$	Generally $\frac{1}{40}$th inch; national map accuracy calls for $\frac{1}{50}$th inch
Adjusted mathematical closure of survey (minimum)	0.03' or 1:50,000	0.05' or 1:50,000	0.07' or 1:50,000	0.1' or 1:50,000	Larger value allowed

TABLE 4   Common Map Symbols

—x—x—	Barbed wire or chain link fence	 ⊘ᴮᴹ	Benchmark
 ⎍ ⎍ ⎍	Wood fence		Corner monumentation
 ▨▨▨▨	Masonry wall	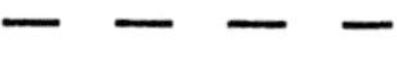 — — — —	Easement
△	Survey control station	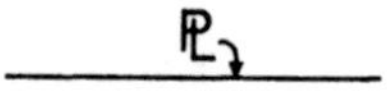 P	Property line
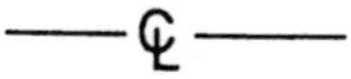 —₵—	Centerline	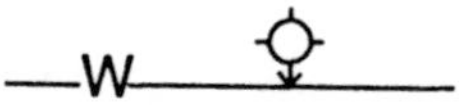 —W— ⊕	Fire hydrant
 —�—	Power line	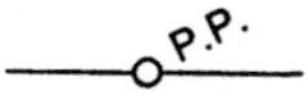 —o P.P.	Power pole
 —W—	Water line	— G —	Gas line
—S—	Sewer line	 — D —	Drain line

TABLE 5   Typical Instrument Precisions

Type of Instrument (Typical)	Smallest Unit Directly Read	Direct Read Uncertainty in 1,000 Feet
Marine compass	11¼ degrees	196.0 feet
Surveyor's compass	15 minutes	4.4 feet
Builder's transit	1 minute	0.29 feet
Mountain transit	30 seconds	0.15 feet
Surveyor's theodolite	10 seconds	0.05 feet[a]
Control survey theodolite	1 second	0.005 feet[a]
Electronic distance meters	0.01 feet	0.03 feet[a]

[a] The errors involved with pointing and centering of target and instrument are much greater than the minimum direct read of the instrument itself.

These standards are presently undergoing revision. Contact ALTA or ACSM for the latest version

Reprinted with permission of the American Congress on Surveying and Mapping.

Minimum Standard Detail Requirements

FOR

ALTA/ACSM
Land Title Surveys

as adopted by

American Land Title Association
and
American Congress On Surveying & Mapping

It is recognized that members of the American Land Title Association (ALTA) have specific problems, peculiar to title insurance matters, which require particular information in detail and exactness for acceptance by title insurance companies when said companies are asked to insure title to land without exceptions as to the many matters which might be discoverable from survey and inspection and not be evidenced by the public records. In the general interest of the public, the surveying profession, title insurers and abstracters, the American Land Title Association and the American Congress on Surveying and Mapping jointly promulgate and set forth such details and criteria for exactness. It is understood that local variations may require local adjustments to suit local situations, and often must be applied. It is recognized equally that title insurance companies are entitled to, and should be able to, rely on the evidence furnished to them being of the appropriate professional quality, both as to completeness and as to accuracy; that it is equally recognized that for the performance of a survey, the surveyor will be provided with appropriate data which can be relied upon in the preparation of the survey.

For a survey of real property and the plat or map of the survey to be acceptable to a title insurance company for purposes of insuring title to said real property free and clear of survey questions (except those questions disclosed by the survey and indicated on the plat or map), certain specific and pertinent information shall be presented for the distinct and clear understanding between the client (in-sured), the title insurance company (insurer), and the surveyor (the person professionally responsible for the survey). These requirements are:

(1) The client, at the time of ordering a survey, shall notify the surveyor that an "ALTA/ACSM LAND TITLE SURVEY" is required, meeting the accuracy requirements of a Class A, B, C, or D Survey as defined in Tables 1 and 2 herein, shall designate which of the additional requirements listed on Table 3 must be included, and shall furnish to the surveyor the record description of the property, documents reflecting any record easements benefitting the property, and the record easements or servitudes and covenants affecting the property ("Record Documents") to which the "ALTA/ACSM LAND TITLE SURVEY" shall subsequently make reference. The names and deed data of all adjacent owners as available, and all pertinent information affecting the property being surveyed, shall be transmitted to the surveyor for notation on the plat or map of the survey.

(2) The plat or map of such survey shall bear the name, address, telephone number, and signature of the professional land surveyor who made the survey, his or her official seal and registration number, the date the survey was completed and the dates of all revisions, and the caption "ALTA/ACSM Land Title Survey" with the certification set forth in paragraph 8.

(3) An "ALTA/ACSM LAND TITLE SURVEY" shall be Class A, B, C, or D, in accor-

dance with the "Classification and Specifications for Cadastral Surveys" as adopted by the American Congress on Surveying and Mapping on March 21, 1986, incorporated herein as Tables 1 and 2. Should these above cited specifications be in conflict with state laws, rules or regulations, the more stringent requirements must be followed.

(4) On the plat or map of an "ALTA/ACSM LAND TITLE SURVEY," the survey boundary shall be drawn to a convenient scale, with that scale clearly indicated. A graphic scale, shown in feet or meters or both, will be included. A north arrow shall be shown and when practicable, the plat or map of survey shall be oriented so that North is at the top of the drawing. If required, supplementary or exaggerated diagrams shall be presented accurately on the plat or map. No plat or map drawing less than the minimum size of 8 1/2 by 11 inches will be acceptable.

(5) The survey shall be performed on the ground and the plat or map of an "ALTA/ACSM LAND TITLE SURVEY" shall contain, in addition to the required items already specified above, the following applicable information:

(a) All data necessary to indicate the mathematical dimensions and relationships of the boundary represented, with angles given directly or by bearings, and with the length of each curve, together with its radius, chord, and chord bearing shown. The point of beginning of the surveyor's description shall be shown as well as the remote point of beginning if different. A bearing base shall refer to some well-fixed bearing line, so that the bearings may be easily re-established. All bearings around the boundary shall read in a clockwise direction wherever possible. The North arrow shall be referenced to its bearing base and should that bearing base differ from record title, that difference shall be noted.

(b) When record bearings or angles or distances differ from measured bearings, angles or distances, both the record and measured bearings, angles, and distances shall be clearly indicated. If the record description fails to form a mathematically closed figure, the surveyor shall so indicate.

(c) Measured and record distances from corners of parcels surveyed to the nearest right-of-way lines of streets in urban or suburban areas, together with recovered lot corners and evidence of lot corners, shall be noted. The distances to the nearest intersecting street shall be indicated and verified. Names and widths of streets and highways abutting the property surveyed and the widths of rights of way shall be given. Any use contrary to the above shall be noted. Access (or lack thereof) to such abutting streets or highways shall be indicated. Private roads shall be so indicated.

(d) The identifying title of all record plats or filed maps which the survey represents, wholly or in part, shall be shown with their filing dates and map numbers, and the lot, block, and section numbers or letters of the surveyed premises. Names of adjoining owners and/or recorded lot or parcel numbers, recording information for last available conveyance, and similar information, where needed, shall be shown. The survey shall indicate set back or building restriction lines which have been platted and recorded in subdivision plats or which appear in a Record Document which has been delivered to the surveyor. Parcel lines shall clearly indicate contiguity, gores, and/or overlaps. Where only a part of a recorded lot or parcel is included in the survey, the balance of the lot or parcel shall be indicated.

(e) All evidence of monuments found or placed, shall be shown and noted to indicate which were found and which were placed. All evidence of monuments found beyond the surveyed premises, on which establishment of the corners of the surveyed premises are dependent, shall be indicated. The character of any and all evidence of possession shall be stated and the location of such evidence carefully given in relation to both the measured boundary lines, as well as those established by the record description. An absence of notation on the survey shall be presumptive of no physical evidence of possession along the record line.

(f) The location of all buildings upon the plot or parcel shall be shown and their locations defined by measurements perpendicular to the boundaries. If there are no buildings erected on the property being surveyed, the plat or map shall bear the statement, "No buildings." Proper street numbers shall be shown where available. All easements evidenced by a Record Document which have been delivered to the surveyor shall be shown, both those burdening and those benefitting the property surveyed, indicating recording information. If such an easement cannot be located, a note to this effect should be included. Observable evidence of easements and/or servitudes of all kinds, such as those created by roads; rights-of-way; water courses; drains; telephone, telegraph, or electric lines; water, sewer, oil or gas pipelines on or across the surveyed property and on adjoining properties if they appear to affect the surveyed property, shall be located and noted. If the surveyor has knowledge of any such easements and/or servitudes, not observable at the time the present survey is made, such lack of observable evidence shall be noted. Surface indications, if any, of underground easements and/or servitudes shall also be shown.

(g) The character and location of all walls, buildings, or fences within two feet of either side of the boundary lines shall be noted. Physical evidence of all encroaching structural appurtenances and projections, such as fire escapes, bay windows, windows and doors that open out, flue pipes, stoops, eaves, cornices, areaways, steps, trim, etc., by or on adjoining property or on abutting streets, on any easement or over setback lines shall be indicated with the extent of such encroachment or projection. If the client wishes to have additional information with regard to appurtenances such as whether or not such appurtenances are independent, division, or party walls and are plumb, the client will assume the responsibility of obtaining such permissions as are necessary for the surveyor to enter upon the properties to make such determinations.

(h) Driveways and alleys on or crossing the property must be shown. Where there is evidence of use by other than the occupants of the property, the surveyor must so indicate on his plan. Where driveways or alleys on adjoining properties encroach, in whole or in part, on the property being surveyed, the surveyor must so indicate on his plans with appropriate measurements.

(i) Cemeteries and burial grounds disclosed in the process of surveying or searching the title to the premises shall be shown by actual location if known.

(j) Ponds, lakes, springs, or rivers bordering on or running through the premises being surveyed shall be shown by actual location.

(k) Streets abutting the premises, which have been legally defined but not physically opened, shall be shown and so noted.

(6) As a minimum requirement, the surveyor shall furnish two sets of prints of the plat or map of survey to the title insurance company or the client. If the plat or map of survey consists of more than one sheet, the sheets shall be numbered, the total number of sheets indicated and match lines be shown on each sheet. The prints shall be on durable and dimensionally stable material of a quality standard acceptable to the title insurance company. At least two copies of the boundary description prepared from the survey shall be similarly furnished by the surveyor and shall be on the face of the plat or map of survey, if practicable, or otherwise attached to and incorporated in the plat or map. Reference to date of the "ALTA/ACSM LAND TITLE SURVEY", surveyor's file number (if any), political subdivision, section, township and range, along with appropriate aliquot parts thereof, and similar information shown on the plat or map of survey shall be included with the boundary description.

(7) Water boundaries are subject to change due to erosion or accretion by tidal action or the flow of rivers and streams. A realignment of water bodies may also occur due to many reasons such as deliberate cutting and filling of bordering lands or by evulsion. Recorded surveys of natural water boundaries are not relied upon by title insurers for location of title.

When a property to be surveyed for title insurance purposes contains a natural water boundary, the surveyor shall measure the location of the boundary according to appropriate surveying methods and note on the plan the date of the measurement and the caveat that the boundary is subject to change due to natural causes and that it may or may not represent the actual location of the limit of title.

(8) When the surveyor has met all of the minimum standard detail requirements for an ALTA/ACSM Land Title Survey, he shall make the following certification on the plat:

To (*name of client*) and (*name of title insurance company, if known*):

This is to certify that this map or plat and the survey on which it is based were made in accordance with "Minimum Standard Detail requirements for ALTA/ACSM Land Title Surveys," jointly established and adopted by ALTA and ACSM in ; meets the accuracy requirements of a Class ___ Survey, as defined therein, and includes Items ___ of Table 3 thereof.

(signed)_____(seal)

Registration No.

Adopted by the Board of Direction, American Congress on Surveying and Mapping on September 16, 1988.

Adopted by the American Land Title Association on October 19, 1988.

American Congress On Surveying and Mapping

Classification and Specifications
For Cadastral Surveys

INTRODUCTION

The degree of precision necessary for a particular cadastral survey should be based on the intended use of the land parcel, without regard to its present use, provided the surveyor has knowledge of the intended use.

Four general survey classes are defined using various state regulations and accepted practices. These general classes are listed and defined in table 1 below.

The combined precision of a survey can be statistically assured by dictating a combination of survey closure and specified procedures for a particular survey class. Table 2 lists the closures and specified procedures to follow in order to assure the combined precision of a particular survey class. The statistical base for these specifications is on file at the ACSM and available for inspection.

TABLE 1

SURVEY CLASSES BY LAND USE

CLASS A—URBAN SURVEYS

Surveys of land lying within or adjoining a City or Town. This would also include the surveys of Commercial and Industrial properties, Condominiums, Townhouses, Apartments and other multiunit developments, regardless of geographic location.

CLASS B—SUBURBAN SURVEYS

Surveys of land lying outside urban areas. This land is used almost exclusively for single family residential use or residential subdivisions.

CLASS C—RURAL SURVEYS

Surveys of land such as farms and other undeveloped land outside the suburban areas which may have a potential for future development.

CLASS D—MOUNTAIN and MARSHLAND SURVEYS

Surveys of lands which normally lie in remote areas with difficult terrain and usually have limited potential for development.

AMERICAN CONGRESS on SURVEYING and MAPPING

TABLE 2

MINIMUM ANGLE, DISTANCE and CLOSURE REQUIREMENTS FOR CLASSES OF SURVEYS
(1)

SURVEY CLASS	DIR. READING OF INSTRUMENT (2)	INSTRUMENT READING ESTIMATED (3)	NUMBER OF OBSERVATIONS PER STATION (4)	SPREAD FROM MEAN OF D&R NOT TO EXCEED (5)	ANGLE CLOSURE WHERE N = NO. OF STATIONS NOT TO EXCEED	LINEAR CLOSURE (6)	DISTANCE MEASUREMENT (7)	MINIMUM LENGTH OF MEASUREMENTS (8), (9), (10)
A	20" <1'> [10"]	5" <0.1'> N.A.	2 D&R	5" <0.1'> [5"]	10" √N	1:15,000	EDM or Doubletape with steel tape	(8) 81m, (9) 153m (10) 20m
B	20" <1'> [10"]	10" <0.1'> N.A.	2 D&R	10" <0.2'> [10"]	15" √N	1:10,000	EDM or steel tape	(8) 54m, (9) 102m (10) 14m
C	[20"] <1'> [20"]	N.A.	1 D&R	[20"] <0.3'> [20"]	20" √N	1:7,500	EDM or steel tape	(8) 40m, (9) 76m (10) 10m
D	[1'] <1'> [1']	N.A.	1 D&R	[30"] <0.5'> [30"]	30" √N	1:5,000	EDM or steel tape	(8) 27m, (9) 51m (10) 7m

Note (1) All requirements of each class must be satisfied in order to qualify for that particular class of survey. The use of a more precise instrument does not change the other requirements, such as number of angles turned, etc.

Note (2) Instrument must have a direct reading of at least the amount specified (not an estimated reading). i.e.: 10" = Micrometer reading theodolite, <1'> = Scale reading theodolite. [10"] = Electronic reading theodolite. (20") = Micrometer reading theodolite, or a vernier reading transit.

Note (3) Instrument must have the capability of allowing an estimated reading below the direct reading to the specified reading.

Note (4) D & R means the Direct and Reverse positions of the instrument telescope, i.e.. Class A requires that two angles in the direct and two angles in the reverse position be measured and meaned.

Note (5) Any angle measured that exceeds the specified amount from the mean must be rejected and the set of angles re-measured.

Note (6) Ratio of closure after angles are balanced and closure calculated.

Note (7) All distance measurements must be made with a properly calibrated EDM or Steel tape, applying atmospheric, temperature, sag, tension, slope, scale factor and sea level corrections as necessary.

Note (8) EDM having an error of 5mm, independent of distance measured (Manufacturers specification)

Note (9) EDM having an error of 10mm, independent of distance measured (Manufacturers specifications)

Note (10) Calibrated steel tape.

TABLE 3

ADDITIONAL SURVEY REQUIREMENTS

If checked, the following additional items shall be shown on the ALTA/ACSM LAND TITLE SURVEY:

1. _____ Monuments placed (or a reference monument) at all major corners of the boundary of the property.
2. _____ Legend of all symbols and abbreviations used.
3. _____ Vicinity map showing the property surveyed in reference to nearby highway(s) or major street intersection(s).
4. _____ Flood zone designation.
5. _____ Land area.
6. _____ Contours.
7. _____ Setback, height and bulk restrictions of record or disclosed by applicable zoning or building codes (in addition to those recorded in subdivision maps). If none, so state.
8. _____ Square footage of all buildings.
9. _____ All improvements (in addition to buildings, such as signs, parking areas or structures, swimming pools, etc.).
10. _____ Parking areas and, if striped, the striping and the number of parking spaces.
11. _____ Indication of access to a public way such as curb cuts, driveways marked.
12. _____ Location of all utilities serving the property, including without limitation:
   (a) all railroad tracks and sidings;
   (b) all manholes, catch basins, valve vaults or other surface indications of subterranean uses;
   (c) all wires and cables (including their function) crossing the surveyed premises, all poles on or within ten feet of the surveyed premises, and the dimensions of all cross wires or overhangs affecting the surveyed premises; and
   (d) all utility company installations on the surveyed premises.
13. _____ Observable evidence of cemeteries.
14. _____ Governmental Agency Requirements:
   Department of Housing and Urban Development
   Veteran's Administration
   Other
15. _____ Significant observations not otherwise disclosed.
16. _____
   _____
   _____
   _____

NOTE: The items of Table 3 must be negotiated between the surveyor and client. It may be necessary for the surveyor to qualify or expand upon the description of these items, e.g. in reference to Item 7, there may be a need for an interpretation of a restriction. The surveyor cannot make a certification on the basis of an interpretation.

APPENDIX B

Reprinted with permission from the U.S. Department of Commerce, National Oceanic and Atmospheric Administration.

TABLE 1.—*Synopsis of horizontal control classifications*

Attributes	Orders of surveys and classes of accuracy				
	Super first-order	First-order	Second-order class I	Second-order class II	Third-order class I, II
General title	Transcontinental control.	Primary horizontal control.	Secondary horizontal control.	Supplemental horizontal control.	Local horizontal control.
Purpose	Transcontinental traverses. Satellite observations. Lunar ranging. Interferometric surveying.	Primary arcs. Metropolitan area surveys. Engineering projects.	Area control. Detailed surveys in areas of very high land value.	Area control. Detailed surveys in areas of high land value.	Area control. Detailed surveys in areas of moderate and low land value.
Network design	Control develops the national network		Control strengthens the national network.	Control contributes to the national network.	Control referenced to the national framework.
Accuracy	1:1,000,000	1:100,000	1:50,000	1:20,000	1:10,000 1:5,000
Spacing	Traverses at 750 km. Spacing—stations at 15 to 30 km or greater. Satellite as required.	Arcs not in excess of 100 km. Stations at 15 km. Metropolitan area control 3-8 km.	Stations at 10 km. Metropolitan area control at 1-2 km	As required	As required.
Examples of use	Positioning and orientation of North American Continent. Continental drift and spreading studies.	Surveys required for primary framework. Crustal movement. Primary metropolitan area control.	Metropolitan area densification. Land subdivision. Basic framework for densification.	Mapping and charting. Land subdivision. Construction.	Local control. Local improvements and developments.

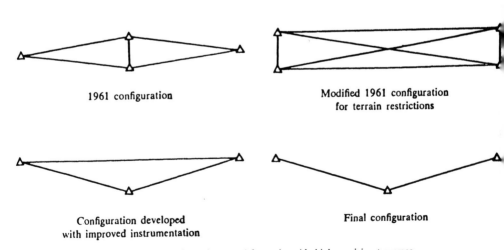

1961 configuration

Modified 1961 configuration for terrain restrictions

Configuration developed with improved instrumentation

Final configuration

FIGURE 1.—Configurations used for nationwide high-precision traverses.

APPENDIX C

Reprinted with permission of the American Congress on Surveying and Mapping.

AMERICAN CONGRESS ON SURVEYING AND MAPPING

Classification and Specifications for Cadastral Surveys

Introduction

The degree of precision necessary for a particular cadastral survey should be based on the intended use of the land parcel, without regard to its present use, provided the surveyor has knowledge of the intended use.

Four general survey classes are defined using various state regulations and accepted practices. These general classes are listed and defined in Table 1 below.

The combined precision of a survey can be statistically assured by dictating a combination of survey closure and specified procedures for a particular survey class. Table 2 lists the closures and specified procedures to follow in order to assure the combined precision of a particular survey class. The statistical base for these specifications is on file at ACSM and available for inspection.

Table 1. Survey Classes by Land Use

CLASS A. Urban Surveys

Surveys of land lying within or adjoining a City or Town. This would also include the surveys of Commercial and Industrial Properties, Condominiums, Townhouses, Apartments, and other multiunit developments, regardless of geographic location.

CLASS B. Suburban Surveys

Surveys of land lying outside urban areas. This land is used almost exclusively for single family residential use or residential subdivisions.

CLASS C. Rural Surveys

Surveys of land such as farms and other undeveloped land outside the suburban areas which may have a potential for future development.

CLASS D. Mountain and Marshland Surveys

Surveys of land which normally lie in remote areas with difficult terrain and usually have limited potential for development.

Table 2. Minimum Angle, Distance and Closure Requirement for Classes of Surveys
(1)

Survey Class	Direct Reading of Instrument (2)	Instrument Reading Estimated (3)	Number of Observations Per Station (4)	Spread from Mean of D&R Not To Exceed (5)	Angle Closure Where N=No. of Stations Not To Exceed	Linear Closure (6)	Distance Measurement (7)	Minimum Length of Measurements (8),(9),(10)
A	20"<1'>[10"]	5"<0.1'>N.A.	2 D&R	5"<0.1'>[5"]	10" $\sqrt{N}$	1:15,000	EDM or double-tape with steel tape	(8) 81 m, (9) 153 m, (10) 20 m
B	20"<1'>[10"]	10"<0.1'>N.A.	2 D&R	10"<0.2'>[10"]	15" $\sqrt{N}$	1:10,000	EDM or steel tape	(8) 54 m, (9) 102 m, (10) 14 m
C	(20")<1'>[20"]	N.A.	1 D&R	(20")<0.3'>[20"]	20" $\sqrt{N}$	1:7,500	EDM or steel tape	(8) 40 m, (9) 76 m, (10) 10 m
D	(1')<1'>[1']	N.A.	1 D&R	(30")<0.5'>[30"]	30" $\sqrt{N}$	1:5,000	EDM or steel tape	(8) 27 m, (9) 51 m, (10) 7 m

Note (1) All requirements of each class must be satisfied in order to qualify for that particular class of survey. The use of a more precise instrument does not change the other requirements, such as number of angles turned, etc.

Note (2) Instrument must have a direct reading of at least the amount specified (not an estimated reading), i.e.:

10" = Micrometer reading theodolite, <1'> = Scale reading theodolite, [10"] = Electronic reading theodolite

(20") = Micrometer reading theodolite, or a vernier reading transit.

Note (3) Instrument must have the capability of allowing an estimated reading below the direct reading to the specified reading.

Note (4) D&R means the Direct and Reverse positions of the instrument telescope, i.e., Class A requires that two angles in the direct and two angles in the reverse position be measured and meaned.

Note (5) Any angle measured that exceeds the specified amount from the mean must be rejected and the set of angles re-measured.

Note (6) Ratio of closure after angles are balanced and closure calculated.

Note (7) All distance measurements must be made with a properly calibrated EDM or steel tape, applying atmospheric, temperature, slope, scale factor, and sea level corrections as necessary.

Note (8) EDM having an error of 5 mm, independent of distance measured (manufacturer's specifications).

Note (9) EDM having an error of 10 mm, independent of distance measured (manufaturer's specifications).

Note (10) Calibrated steel tape.

GLOSSARY OF SELECT TERMS

Abstract of title (or deed)   A summary of conveyances, exchanges, easements, or other legal instruments affecting the property rights of a particular parcel.

Accessories to a corner   Natural physical objects, such as trees, rock formations, ledges, and other features, that are at a known distance and/or direction from a corner. Accessories are part of the corner monumentation.

Adjoiner   That parcel of land that shares a common boundary with another.

Adjusted value   The value assigned to a measured quantity after the application of corrections designed to account for measurement errors.

Aliquot parts   A division of a USPLS section following a specified procedure.

Angle   A measure of the relationship of two intersecting lines.

Area   The measure of the bounded surface formed by the intersection of real property boundaries and a particular vertical datum.

Astronomic   Values assigned to direction or position based upon measurements made of the relative positions of heavenly bodies.

Azimuth   A definition of the direction of a line based upon the clockwise angle formed between that line and a certain pole of a meridian.

Balancing a traverse   A procedure, or procedures, for distributing the accumulated measurement errors of a traverse among the observed values in order to obtain computational consistency.

Baseline   A series of points established for the expressed purpose of locating other features or lines.

Bearing   A definition of the direction of a line based upon the clockwise or counterclockwise angle formed between that line and either pole of a meridian.

Benchmark   An object, natural or artificial, in a relatively stable location, that is at a known elevation relative to a particular vertical datum.

Blunder   A mistake or an incorrect assessment of a measured value associated with a gross misinterpretation of facts.

Boundary line   An imaginary line of demarcation between two adjoining land parcels, distinguishing a separation of real property rights, which may, or may not, be physically marked.

Bureau of Land Management (BLM)   An agency of the U.S. government, formerly known as the Government Land Office (GLO), which is responsible for the survey of public lands, among other things.

Cadastre   An official map of a political region delineating size, location, ownership, and land values for the purpose of assessing taxes.

Center of a section   A point in a USPLS section determined by the intersection of two straight lines drawn between opposing quarter corners.

Chain   (1) A length of measure equal to 66 U.S. survey feet; (2) a surveyor's measuring tape; (3) the act of measuring a linear distance.

Chord   A straight line drawn between the ends of a curved line segment.

Compass   (1) A device for detecting the earth's magnetic field and aligning with the lines of force; (2) a device for laying out a specific distance or marking an arc; (3) a unit of measure, usually defined as 6 feet.

Contour   A series of lines on a map connecting points of equal elevation.

Control   A series of vertical and/or horizontal survey marks, data, measurements, maps, photographs, or other acts specifically performed to serve as the foundation or datum for future surveys.

Coordinate system   A method of identifying a particular point in two or three dimensions by a systematic listing of the distances from defined baselines or origins.

Corner   A point of intersection of real property boundary lines, which may, or may not, be monumented.

Course   The direction of a line segment. In some states, the course also includes the length of a line segment.

Datum   A basis or measurement foundation on which a location can be defined or referenced either vertically, horizontally, or both.

Deed   A written instrument that conveys rights or interests in real property.

Deed description   That part of a deed that identifies and describes the relevant real property parcel.

Departure   The change in the distance from a meridian experienced in moving from one end of a line segment to another.

Easement   A right held by one party to the land of another.

Electronic distance meter (EDM)   A device that measures distances by the use of electromagnetic radiation.

Elevation   The distance, above or below, a vertical datum.

Equator   (1) Zero degrees latitude; (2) an imaginary plane passing through the earth and perpendicular to the axis of rotation.

Error   The difference between the measured and actual values for a certain quantity.

Error of closure   The failure of the result of a set of measured values to agree mathematically with the theoretical result.

Fee simple   Unlimited rights of real property ownership restricted only by the laws of the U.S., state, and local governmental jurisdictions.

Field notes   The written notes, sketches, and computations of a surveyor taken during and at the site of a survey.

Geoid   A theoretical surface that is everywhere perpendicular to plumb at every point.

Government Land Office   See Bureau of Land Management (BLM).

Grade   The slope of a surface or structure.

Grid   An Imaginary network of evenly spaced parallel and perpendicular two-dimensional lines.

Horizontal   A plane, perpendicular to the plumb line at a particular point.

Land description   The exact location of a real property parcel in terms of the controlling land record system.

Legal description   That description of a real property parcel sufficient to identify that parcel uniquely without oral testimony.

Latitude   (1) The distance along a meridian; (2) the north or south change in distance experienced in moving from one end of a line segment to another.

Level   (1) A surface that is everywhere perpendicular to plumb; (2) an instrument for measuring differences in elevation.

Line   A series of contiguous points forming a vertical geometric surface and extending from the center of the earth up through the land surface and into space.

Longitude   The distance between two meridians.

Map A graphic, two-dimensional representation of the surface of the earth.

Map projection A systematic method, accounting for the curvature of the earth, that mathematically reduces surface location information into two-dimensional data.

Mean sea level (MSL) The average elevation of the sea over a 19-year period. MSL is often confused with the National Geodetic Vertical Datum (NGVD).

Measurement An estimation of a quantity or a distance based upon the systematic application of a standardized procedure or device.

Metes and boundsdescription A description formed by sequentially reciting the courses and adjoiners of a real property parcel.

Meridian A north–south line used to reference lines of a survey.

Monument The physical object that indicates the location of a point, station, or real property corner.

More or less A phrase indicating a crude or uncertain value for a quantity.

National Geodetic Survey (NGS) The agency of the U.S. government that is responsible for development and maintenance of benchmarks and stations for navigation and mapping.

National Geodetic Vertical Datum (NGVD) The vertical datum established by the National Geodetic Survey that defines elevations published for use on federal maps and regulations.

North Aligned with the axis of the earth's rotation and in the direction of that particular pole designated as "north."

Observation The noting of a condition, which usually is the result of an act of measurement.

Plane A flat surface such that the shortest route between any two locations on the surface is entirely contained within the surface.

Plat A map, prepared by a land surveyor, usually for a specific legal purpose.

Plumb Aligned with the pull of gravity.

Platted subdivision description A description based upon a map or plan, usually recorded, identifying a real property parcel by the letter or number designation found on that map or plan.

Point (1) A specific location; (2) a vertical geometric ray (line) originating at the center of the earth and extending up through the surface into space.

Point-of-beginning The first point encountered in the narrative portion of a deed description, especially a metes and bounds description, that is a part of the real property boundary itself.

Point of compound curvature   A point on a circular curved line at which a curve of one radius length ends and a curve of another radius length begins, and occurring at a point on the extension of a straight line drawn between the two radius centers.

Point of curvature   A point at which a straight line begins a circular curve and is at right angles to the radius of the curve at that point.

Point of reverse curvature   A point on a curved line, occurring on a line drawn between the radius centers, at which a curve in one direction ends and a curve in the opposite direction begins.

Pole   *See* Rod.

Quarter corner   That corner, set by the government surveyor, between section corners.

Random error   Incidental errors occurring as a result of observational imprecision.

Random traverse   A traverse in which the location of stations is chosen for accessibility and intervisibility and does not have a constant relationship to any real property boundaries.

Range line   A north–south line used to divide public lands.

Recovered corner   A real property corner that has been verified by the discovery of the original monument, accessories, or other physical evidence.

Right-of-way   Land granted (usually to the governing authority) by deed, servitude, or easement for the construction of an infrastructure. Rights-of-way may grant limited property rights or full property rights.

Riparian   Pertaining to the banks of a body of water.

Rod   (1) $16\frac{1}{2}$ feet; (2) a wooden pole (rod) being $16\frac{1}{2}$ feet long and used to measure horizontal distances.

Section   The smallest division of land monumented by the U.S. government when it subdivides public lands for sale.

Servitude   *See* Easement.

Spiral curve   A curved line of constantly varying radius.

Station   A point established or measured by survey procedures.

Straight line   A vertical plane containing the center of the earth. A line established by the line of sight between two points.

Surface   The separation of two distinct spaces. The interface between the earth and the atmosphere.

Land surveying   The art and science of measuring, marking, recovering, and mapping the relative positions or locations of terrain features and real property boundaries.

Systematic error   Errors occurring consistently, regularly, and of the same algebraic sign as the result of a measurement condition.

Theodolite   An instrument designed to measure precisely vertical and horizontal angles.

Title   The exclusive right to the use and enjoyment of a particular parcel of real property.

Township   A division of public lands generally 6 miles wide and 6 miles long and containing 36 sections.

Township line   An east–west line used to divide public lands.

Transit   In the United States, a transit is a theodolite having a vernier read scale for measuring horizontal and vertical angles and having a scope that is capable of being inverted.

Traverse   A systematic series of stations in which the direction and length of line segments formed by consecutive stations is measured.

Vernier   An etched ruler or scale that is marked such that, when it is aligned with another ruler or scale, divisions much smaller than marked on either scale can be read directly.

Vertical   Aligned with the pull of gravity.

Witness mark   A monument that is at a known distance and direction from a corner.

BIBLIOGRAPHY

Those readers interested in a more detailed study of the various scientific and legal points touched on in this book are referred to the following list of excellent works.

American Congress on Surveying and Mapping. *Legal Topics in Boundary Surveying: a Compendium,* 1990

American Congress on Surveying and Mapping and the American Society of Civil Engineers. *Definitions of Surveying and Associated Terms,* 1978.

Brinker, Russell C., and Roy Minnick. *The Surveying Handbook.* New York: Van Nostrand Reinhold Company, 1987.

Brown, Curtis M., Walter G. Robillard, and Donald A. Wilson. *Boundary Control and Legal Principles,* 3rd ed. New York: John Wiley & Sons, 1986.

————, Walter G. Robillard, and Donald A. Wilson. *Evidence and Procedures for Boundary Location,* 2nd ed. New York: John Wiley & Sons, 1981.

Buckner, R. B. *A Manual on Astronomic and Grid Azimuth.* Rancho Cordova, CA: Landmark Enterprises, 1984.

————. *Surveying Measurements and Their Analysis.* Rancho Cordova, CA: Landmark Enterprises, 1983.

Clevenger, Shobal V. *A Treatise on the Method of Government Surveying.* New York: D. Van Nostrand, 1883.

Davis, Raymond E., Francis S. Foote, and Joe W. Kelly. *Surveying: Theory and Practice,* 5th ed. New York: McGraw-Hill Book Company, 1966.

Kiely, Edmond R. *Surveying Instruments: Their History.* Columbus, OH: Carben Surveying Reprints, 1979.

Kissam, Philip. *Surveying for Civil Engineers.* New York: McGraw-Hill Book Company, 1956.

Mackie, J. B. *The Elements of Astronomy for Surveyors.* London and High Wycombe: Charles Griffin & Company, Ltd., 1978.

McEntyre, John G. *Land Survey Systems.* New York: John Wiley & Sons, 1978.

Mikhail, Edward M., and Gordon Gracie. *Analysis and Adjustment of Survey Measurements.* New York: Van Nostrand Reinhold Company, 1981.

Mulford, A. C. *Boundaries and Landmarks.* New York: D. Van Nostrand, 1912.

Robillard, Walter G., and Lane J. Bouman. *Clark on Surveying and Boundaries.* Charlottesville, VA: The Michie Company, 1987.

Skelton, Ray Hamilton. *The Legal Elements of Boundaries and Adjacent Properties.* Indianapolis, IN: The Bobbs-Merrill Company, 1930.

Stewart, Lowell O. *Public Land Surveys.* Minneapolis, MN: The Meyers Printing Company, 1935.

U.S. Department of Commerce, National Oceanic and Atmospheric Administration. *Specifications to Support Classifications, Standards of Accuracy, and General Specifications of Geodetic Control Surveys.* Rockville, MD: 1980.

U.S. Department of the Interior, Bureau of Land Management. *Restoration of Lost or Obliterated Corners and Subdivision of Sections.* Washington, D.C. U.S. Government Printing Office, 1979.

Wattles, William C. *Land Survey Descriptions.* Orange, CA: Parker & Son, Inc., 1974.

Wilson, Donald A. *Deed Descriptions I Have Known . . . but Could Have Done Without.* Rancho Cordova, CA: Landmark Enterprises, 1982.

INDEX